PENGUII

JUSTINIA

William Rosen was a senior executive a ......ui ana Simon & Schuster publishing houses for more than twenty-five years. He lives in Princeton, New Jersey. *Justinian's Flea* is his first book.

Praise for *Justinian's Flea*

"This is a remarkable book; the polymath author expatiates on subjects as varied as architecture, theology, jurisprudence, the art of war, and, of course, biology, as well as happily imparting the juiciest bits of imperial gossip. Part III, Bacterium, begins with a detailed account of the nature of the plague, traces its terrible spread across ancient Europe, and provides visceral descriptions of the ravaged Constantinople, 5,000 citizens dying every day. The end of the world, thanks to a flea."

—Talha Burki, *The Lancet*

"This is a very engaging, lively, and entertaining text that presents the story of the plague almost as a mystery tour. . . . An impressively wide-ranging book covering epidemiology, medical history, economics, agricultural history, evolution, and architecture. Rosen marshals information from history, zoology, genetics, complexity theory, meteorology, and evolutionary history to present a fascinating account of the interrelationships between fleas, rats, bacteria, climate, and food supply. . . . Worth reading for the lively accounts of battle scenes and the movement of armies; sympathetic descriptions of early Christian debates about the nature of God and Christ alongside critiques of intelligent design; discussions of the structure of marble; explanations of the effects of bacteria on a flea's stomach and appetite, and, ultimately, the effect all this had on an empire."

—Helen Blackman, *The Journal of the American Medical Association*

"Rosen's knowledge of these events is remarkable. His explanations of the interaction of *Y. pestis* with the immune system and of bacterial pathogenesis provide insight into a killer that shaped history. . . . *Justinian's Flea* is a well-researched book that is also a pleasure to read, and I enthusiastically recommend it."

—Raymond J. Dattwyler, M.D., *The New England Journal of Medicine*

JUSTINIAN'S FLEA

THE FIRST GREAT PLAGUE
AND THE END OF
THE ROMAN EMPIRE

WILLIAM ROSEN

PENGUIN BOOKS

PENGUIN BOOKS

Published by the Penguin Group

Penguin Group (USA) Inc., 375 Hudson Street, New York, New York 10014, U.S.A.

Penguin Group (Canada), 90 Eglinton Avenue East, Suite 700, Toronto,
Ontario, Canada M4P 2Y3 (a division of Pearson Penguin Canada Inc.)

Penguin Books Ltd, 80 Strand, London WC2R 0RL, England

Penguin Ireland, 25 St Stephen's Green, Dublin 2, Ireland (a division of Penguin Books Ltd)

Penguin Group (Australia), 250 Camberwell Road, Camberwell,
Victoria 3124, Australia (a division of Pearson Australia Group Pty Ltd)

Penguin Books India Pvt Ltd, 11 Community Centre,
Panchsheel Park, New Delhi – 110 017, India

Penguin Group (NZ), 67 Apollo Drive, Rosedale, North Shore 0632,
New Zealand (a division of Pearson New Zealand Ltd)

Penguin Books (South Africa) (Pty) Ltd, 24 Sturdee Avenue,
Rosebank, Johannesburg 2196, South Africa

Penguin Books Ltd, Registered Offices:
80 Strand, London WC2R 0RL, England

First published in the United States of America by Viking Penguin,
a member of Penguin Group (USA) Inc. 2007
Published in Penguin Books 2008

10 9 8 7 6 5 4 3 2 1

Copyright © William Rosen, 2007
All rights reserved

ISBN 978-0-670-03855-8 (hc.)
ISBN 978-0-14-311381-2 (pbk.)
CIP data available

Printed in the United States of America
Designed by Carla Bolte · Set in Adobe Garamond
Maps by David Lindroth

FOR JEANINE

*The two greatest problems in history
are how to account for the rise of Rome,
and how to account for her fall.*

—ERNEST RENAN

CONTENTS

PART IV: PANDEMIC

MAPS

JUSTINIAN'S FLEA

The Three Thousand-Body Problem

T HE LAW OF gravitation discovered in the seventeenth century by Isaac Newton states that two bodies attract each other with a force that is directly proportional to the centers of the bodies' mass, and inversely proportional to the square of the length of a straight line separating one from the other. What this means, in practice, is that the measure of the path followed by any two bodies—the earth and the moon, for example—has a single solution, depending on the size of the respective bodies, and their distance from one another. Solutions to such two-body problems, as they are known, are simple, elegant, and most of all, unique.

Add even one more body to the system, however, and the solution is no longer unique. The best solutions to three-or-more-body problems, in fact, are "only" approximations . . . though, with the help of powerful computers, those approximations can be extremely precise. Calculating the path of an Apollo spacecraft from the east coast of Florida to the Sea of Tranquility, for example, which must take into account the mass of the earth, the moon, the sun, as well as the spacecraft itself (not to mention relatively minor effects exerted by other planets, comets, stars, and so on), is considerably harder than figuring the path of a dozen billiard balls on a felt tabletop, but the approximate solution has clearly been a satisfactory one.

The forces that transformed the Mediterranean world of late antiquity into medieval Europe were considerably greater in number than the significant gravitational forces acting on *Apollo 11,* and any history that proposes a precise account of their interactions is bound to be, in some respects, unsatisfactory. These forces include, in no particular order, the geography and climate of the Mediterranean and surrounding territories; the eastward shift of the Roman empire from its Italian home to Asia Minor; the resulting westbound migrations of numerous peoples—preeminently the Goths—from the Black Sea region into the Italian and Iberian peninsulas; and the encounter with successive waves of nomadic horse archers emerging out of the Eurasian steppe (whose presence initiated a series of military innovations that led to the armed, armored, and stirruped cavalry of the European Middle Ages). Other forces, no less powerful, acted directly on the minds of the peoples of the Mediterranean: the centuries-in-refinement philosophy of Plato, particularly as seen in the late antiquity development retrospectively known as Neoplatonism; messianic Christianity, with all its attendant and perhaps inevitable, doctrinal disputes; and the growth of powerful educational institutions drawing on both traditions, whose pupils rose to the highest positions in government.

Some of the forces were initiated by individuals: the military revolution of Diocletian, the adoption of Christianity as a state religion by Constantine, and the reenergizing of the Persian Empire—and *its* state religion, Zoroastrianism—by its great ruler, Khusro Anushirvan all changed the path of history to greater or lesser degree. Some of the forces exerted influence in the most unlikely ways; one of the consequences of the theft of the secret of silk making from China was the withdrawal of Rome and Persia from the Arabian peninsula, only a decade before the birth in Mecca of the founder of the world's last great monotheistic religion . . . one whose armies would destroy the Persian Empire and conquer most of Rome's.

The orbital path of an object in a two-body system, or, at least, one with only two gravitationally significant bodies, is a regular one: an ellipse. To know the position and speed of such an object at a specific moment in time makes it possible to determine its position at any subsequent moment. A three-body problem—call it a three thousand-body problem—is different. In a three thousand-body system like the sixth-century Mediterranean, even knowing the status of a given object at two such moments—the first dominated by Rome and Persia, Plato and Christ, Athens and Jerusalem; the second by Islam and Chris-

tendom, Muhammad and Aquinas, Baghdad and Cologne—tells little
of the route that the object followed to get from one to the other.

Though a precise retracing of the journey is impossible, however,
some approximations are better than others, else the writing of history
itself would be impossible. It is not necessary to argue that any single
phenomenon gave birth to the nation-states of Europe to find merit
in examining the moment of their conception. This moment, a hinge
time if ever there was one, has historically been the century that occupies the last chapter in books on the classical world, or the world of
antiquity . . . or, contrariwise, the first chapter in histories of the medieval world. It is the century, and the moment, when the last Roman
emperor who deserves to be called great embarked on the reconquest
of Italy, Spain, and North Africa; put his entire weight behind the reconciliation of schism in the Christian episcopate; and drew on the esthetic and intellectual capital of the great achievements of both the
eastern and western Mediterranean world to build some of the world's
greatest architectural *and* legal edifices.

And it is the moment, with the emperor at the absolute zenith of his
achievement, that the world encountered the first pandemic in history.

The coincidence of timing does not, of course, prove that the pandemic caused Rome to fall, or Europe to be born; as above, the uncer-
tainties of the three thousand-body problem makes such a claim
fundamentally uncertain. However, the Plague of Justinian, to give
both pandemic and emperor their names, killed at least twenty-five
million people; depopulated entire cities; and depressed birth rates for
generations precisely at the time that Justinian's armies had returned
the entire western Mediterranean to imperial control and only decades before Muhammad's followers emerged out of Arabia to conquer Egypt, Palestine, Syria, Libya, Persia, Mesopotamia, and Spain. It
is therefore as difficult to plot a course to modern Europe without ac-
knowledging the presence of Justinian and the plague as it would be
to send a satellite to the moons of Saturn without accounting for the
gravitational impact (the technical word is *perturbation*) of the planet
Jupiter.

Gravitational perturbation has proved a most useful tool to astronomers, most famously in 1846, when some twitchiness in the orbit of the visible planet Uranus was ascribed to the gravitational mass of the not yet visible planet Neptune. So, too with human history. When two nations share a similiar historical position at the same point in time, and their paths subsequently diverge, it's worth searching for some powerful if unseen gravitational mass. Consider the cases of Rome and China.

During the fourth century of the Common Era, both great Eurasian empires were threatened with permanent dissolution. The empires founded by Augustus in 31 B.C.E. and Shih huang-ti in 221 B.C.E. each faced invasion by peoples termed "barbarians" and had, as a result, suffered significant reductions in imperial size, wealth, and even legitimacy . . . so much so that by the sixth century, the respective rulers of each empire embarked on a planned *reconquista*.

And each emperor succeeded, at least initially. In China, Yang Chien was successful in reasserting imperial authority over that portion of northern China that had, two centuries earlier, fallen to the barbarians known to historians as the Sixteen Kingdoms. And, from his capital in Asia Minor, an unlikely place from which to rule an empire still called Roman, Justinian the Great defeated, in order, the Vandals, Ostrogoths, and Visigoths, reconquering North Africa, Italy, and a fair bit of Spain.

But while the first acts of the Chinese and Roman reunification dramas were similar, their denouements were wildly different. The T'ang dynasty, which succeeded the Sui dynasty founded by Yang Chien, reigned over a united China until the tenth century. Over the same period of time, Justinian's successors were not only unable to maintain imperial authority over Italy and Spain but also lost Egypt, Syria, the Balkans, North Africa, and Mesopotamia. While the T'ang emperors ruled all of China, the Roman Emperor—he was still called nothing else—ruled over little more than Greece and the city of Constantinople. The modern world would be dominated by the technological and military ambitions of the European nations that replaced the Roman superstate.

Why did China not atomize into half a dozen different kingdoms separated by language, as happened in Europe? The differences between the two are, of course, massive, from geography to technological developments to religious history, and most of them are far beyond the scope of this book. The simplest explanation, during the period when Rome dissolved and China coalesced, was the growth of the Islamic Caliphates, against which Christendom (as Europe was then known) was obliged to define itself. (Though, as H. L. Mencken supposedly said, "For every complex problem there is a simple solution . . . and it is wrong.") Since even the prospect of a powerful enemy was insufficient to unite Europe, insufficient to keep Spain and England and France from finding their separate national identities, the forces of atomization must have been even more powerful. The obvious inference is that a common sovereign is a far more effective force for union than a common enemy, and the absence of that sovereign—the absence of Imperial Rome—permitted the pieces of Europe to define themselves based on local, rather than universal, characteristics.

The act of definition continues today. The modern political entity known as Europe—the one attempting, with limited success as of this writing, to establish a continent-wide legal charter—is a creation of imagination as much as one of geography. More so, in fact; the geography of "Europe" is a matter that has confounded historians from Herodotus to the present. In at least three different places in his great history of Justinian's reign, Procopius attempts to define Europe by its geographical boundaries . . . and, like everyone else, runs aground in the east. The decision to separate Europe from Africa (Procopius called it Libya) at Gibraltar is both historically and geographically obvious. The line separating Africa from Asia at the Nile is at least defensible (though problematic, given the presence of Egyptians on both banks of the boundary for millennia). But selecting the boundary line between Europe and Asia is daunting on both geographical and historical grounds; Procopius offers two rivers feeding into the Black Sea: the Tanais (Don) and Phasis (Rioni) rivers. The Phasis appeals not only because of its topography but because of its appearance in the foundational myth of European-Asian conflict, that of Jason's search for the Golden Fleece, the preface to the Trojan War. The modern bureaucrats wrestling with the possibility of a European Union containing both Turkey and Ukraine walk in the steps of historians dead for fifteen centuries.

Even for Procopius and his contemporaries, Europe had been a problematic idea for millennia. The earliest appearance of the term is literally mythic: Zeus's abduction of Europa from her home in her father's kingdom of Tyre, with her subsequent rape and abandonment on the shores of the European continent. (Herodotus's version has Cretan merchants abducting her for their king Asterius[1]) Subsequent myths defined Europe as a people engaged in armed conflict with "Asiatics," beginning with Thermopylae and climaxing retrospectively in Virgil's epic of the departure of Aeneas from Asia to found Rome, which in turn founded "Europe."

Both the "Europe" of myth and the Europe of the Greco-Roman world were really the world of the Mediterranean, the center of the world to historians as diverse as Fernand Braudel, G. F. W. Hegel, and Henri Pirenne, who was the first to argue, in *Mohammed and Charlemagne*, that the end of antiquity was marked by the Islamic invasions of the seventh century . . . invasions that are a direct consequence of the events related in *Justinian's Flea*. The thousand-year-long transformation of a Mediterranean superstate into a northern European collection of nation-states by the time of the Treaty of Westphalia in 1648, may seem, in retrospect, inevitable. It is anything but. The idea of political entities, supreme within their own borders, entitled to something called sovereignty and defined by geography, language, and common descent, is not an exclusively European creation, but nearly so. Such nation-states did not—could not—come into being so long as the dominant political structure of the day was Rome's empire, membership in which depended not upon birthplace or birth parents, but on acceptance of Roman authority and Roman law.

A thousand threads connect "Rome" with "Europe," and *Justinian's Flea* is therefore a book of connections: an attempt to place in context a moment in history by weaving a tapestry out of the threads that connect it to the world that replaced it. Some of those threads are technological, some military, some geographic. They include evolutionary microbiology, architecture, animal ecology, jurisprudence, theology, and even commerce. To follow any of these threads is to take a journey out of one world and into another.

These different threads, however, can be easier to follow than to weave. The goal of any history, it seems to me, is to present a picture that embraces both the exotic and the familiar—that shows a moment in time as both similar to the reader's own, and fundamentally different. To accomplish this with Justinian's world requires both an account of the forces that *affected* the late Roman empire—impersonal trends like demography and climate, human creations from armies to ideas—and the forces that the late empire *exerted* on the European nations that succeeded it. Doing justice to both means appreciating the historical circumstances into which Justinian was born, the legacy he left behind, and the monumental and contingent accidents that accompanied him along his journey.

Consequently the structure of this book—the warp and weft of the tapestry promised above—is divided into four parts. Part I, "Emperor," describes the westward "barbarian" migrations as a reaction to the great eastward move of the newly Christian empire from Rome to Constantinople, each pivoting on the Balkans, the birthplace of Constantine, Diocletian, and Justinian. "Glory," Part II, is an account of Justinian's triumphs as they influenced the formation of the nations, the law, and the architecture of medieval Europe, and is therefore an appreciation of his contributions as a conqueror, as a jurist, and as a builder. In Part III, "Bacterium," the other great player in the drama of the end of late antiquity appears: bubonic plague, its evolution, and its horrific impact. "Pandemic," the fourth part, follows the journey of the plague to Persia, France, Britain, and Italy, and, equally important, visits the territories where the plague's absence had an equally profound effect: China and Arabia.

This book began with a belief that the European history that occupies such a large part of the knowledge of every educated person was a contingent history, one that could have taken some very different turns. Nothing learned in the course of its writing has suggested precisely where those turns might have taken Europe, but it has taught a new respect for the Roman world that preceded it. Justinian's reign, typically, displays the well-known Roman genius for soldiering and administration; it also, perhaps instructively, is an object lesson in how the world's greatest military power can find military victories undone by local insurgencies. But in some ways, the Roman quality that appeared most admirable was a surprising one: Roman openness.

It should not, in retrospect, have been so surprising to discover that Rome was such a socially mobile place. During the period of this book, the Roman Empire was ruled by an emperor who was himself born a peasant, and an empress who was a onetime courtesan, neither of whom ever set foot in Rome itself, or even Italy.

Examples of the assimilative genius of Rome still have the power to surprise; of the great nations of history, only three—significantly, given the compass of this book, the first two are Imperial China (Li Po, the greatest poet of T'ang China, was an ethnic Turk) and Golden Age Islam; the other notwithstanding a robust tradition of nativism, is the United States of America—rival Rome in ease of entry. One cannot simply decide to become German, or Russian, or Korean in the way that one could become Roman. But even slaves could, and frequently did, become Romans, since Rome vested the right of manumission in every slave owner, and defined it as not merely freedom, but citizenship. While imperial Rome—autocratic, militaristic, arrogant Rome—is scarcely the Kingdom of Heaven, given the horrors of the *blut-und-böden* states that replaced it—Crusades, Inquisition, Holocaust, Gulag—one might, perhaps, be forgiven some wistfulness at its passing.

And, perhaps, some wonder at the possibility that that passing was hastened by the bite of a flea.

PROLOGUE

Pelusium

540

During these times, there was a pestilence, by which the whole human race came near to being annihilated. Now in the case of all other scourges sent from heaven some explanation of a cause might be given by daring men, such as the many theories propounded by those who are clever in these matters, for they love to conjure up causes which are absolutely incomprehensible to man. . . . But for this calamity, it is quite impossible either to express in words or to conceive in thought any explanation, except indeed to refer it to God . . .

It started from the Egyptians who dwell in Pelusium.

—Procopius, *History of the Wars*, II, xxii

B Y THE MIDDLE *of the sixth century Pelusium was more than a thousand years old, a fortress town built at the mouth of the easternmost branch of the Nile by the Persians on the site of their victory over the Egyptians in 525 B.C.E. The place was old even then. Under its original Egyptian name, Saʾina—in English, the Wilderness of Sin—it appears in the second book of the Old Testament at both the beginning and the end of the Exodus, first as the site where God gave the Israelites manna for their hunger, and last as the place from which Joshua sent his spies into Canaan. Pelusium had later been conquered by Julius Caesar, and watched the escape of Hannibal to Rhodes. The frontier city had seen Pontius Pilate sailing east on his way to Judea, and the Jews going west on their journey into exile.*

But it had never witnessed anything like the corpses. Not, to be sure, that dead bodies were a novelty in a time when a man who lived past his fortieth year was considered fortunate, and one child in four never celebrated a first birthday. Disease, even mortal disease, was nothing new, in Egypt or anywhere else. Even so, the corpses did have a distinctive feature, the grapefruit-sized swellings in groin and armpits that were called, in Greek, buboes.

The disease that appeared in Pelusium in the 540th year of the Common Era was lethal—seven in ten victims died within a week—but it was containable. Like a house fire in the middle of a desert, and in fact like every earlier disease outbreak in human history, the pestilence would soon burn itself out for lack of fuel . . . or, it would have but for one thing. One hundred and sixty miles to the west of Pelusium, on the other side of the Nile delta, was its much larger cousin, Alexandria. Its proximity had enormous consequence. Once the pestilence migrated across the delta, it found a huge new fuel source in the hundreds of thousands of residents of the Mediterranean's second largest city. And, far more dangerously, it found the ships.

The ships that entered and departed the harbor of Alexandria each day were of all types. Some were galleys, moved by banks of oars. Others were wind-powered, sometimes with square sails, sometimes rigged with the triangular lateen that could be used to sail before the wind. There were tiny boats that displaced less than fifteen tons used for the coasting trade, grain ships of several hundred tons, and giants of more than a thousand tons, specially built to carry the giant obelisks used for monuments.[1] The ships of Alexandria knit the world's greatest empire together, carrying grain from Egypt to Apulia, and copper from Cyprus to Spain. A bishop in Gaul could anoint a novitiate with oil from the olive orchards of Greece, bless the event with wine from the vineyards of Italy, and celebrate the sacrament with bread baked with the wheat of Africa while wearing a garment made by Syrian weavers from Chinese silk, all because of the ships.

In the year 540, the ships that meant life to the great trading cities of the Mediterranean littoral left Alexandria as they had for centuries, carrying freight, crew . . . and rats, which carried a cargo of their own.

———

On a morning in the spring of 542, on the opposite shore of the Mediterranean from Pelusium, the most powerful man in the world greeted the dawn. Ruling the world's richest and greatest empire was a job without end, and so it had become his custom to work straight through the night, foregoing food and sleep. The man's work habits had convinced his enemies that he was not a man at all, but a demon requiring no rest. They whispered that he was a headless monster that prowled the halls of his palace and the streets of his city while honest people slept, a demon whose appetite demanded more souls than there are grains of sand on the beach. The ruler's ambition had carried him from a Balkan peasant village to the peak of the known world, and was not yet satisfied. Nor would it be until his domain had regained the extent enjoyed under his predecessors.

Within months, the ships would arrive, carrying a real demon. The collision between demon and peasant-turned-ruler would mark the end of one world, and the beginning of another. Along the way, it would consume at least twenty-five million human lives.

PART I

EMPEROR

CHAPTER ONE

"Four Princes of the World"

286–470

> The people Romans call, the city Rome.
> To them no bounds of empire I assign,
> Nor term of years to their immortal line.
> Ev'n haughty Juno, who, with endless broils,
> Earth, seas, and heav'n, and Jove himself turmoils;
> At length aton'd, her friendly pow'r shall join,
> To cherish and advance the Trojan line.
> The subject world shall Rome's dominion own,
> And, prostrate, shall adore the nation of the gown.[1]
> —Virgil, *The Aeneid*

THE BALKAN HILL town of Tauresium appears on no modern atlas, and was almost certainly absent from maps that were in use during the centuries that modern historians call late antiquity. The only reason that the village, in the Roman province of Illyricum, is remembered today is that, in the closing years of the fifth century of the Common Era, a boy departed it. Twelve years earlier, his mother, a peasant girl named Vigilantia, had christened him Petrus Sabbatius. Many years later, after his journey to the capital of what was still the world's largest empire, he was known by the name he gave himself: Justinian.

The journey of Petrus Sabbatius mirrored that of the empire he was eventually to rule. Two hundred years before the boy left his village, probably somewhere near modern Skopje, in Serbia-Macedonia, the empire itself had begun a centuries-long migration eastward. Before that, horsemen from the great grassland that stretches all the way to the mountains of China had arrived, heading west. Soldiers and farmers, priests and merchants, Christians and pagans had marched and countermarched across the lands north of the Mediterranean for centuries, and all their journeys, sooner or later, passed through the Balkans, and Illyricum.

By the end of the fifth century the Illyrians, who gave the province its name, had been living along the Dalmatian coast and the mountains of the Balkan interior for nearly two thousand years. Their villages and towns were the latest stratum of cultural sediment dating back two hundred thousand years, to the early Paleolithic. An auger sunk deeply enough into the Balkan soil at the time of Justinian's birth would have revealed a tiered core sample of Neolithic ceramics, Bronze Age stone forts, and Greek statuary left behind from fourth-century B.C.E. colonies. The top layer of the sandwich would have recorded the previous five hundred years, the five centuries of rule by Rome's emperors.

Roman influence dated even further back, to the days of the Republic; a portion of the surrounding area had been conquered by the legions in the year 168 B.C.E., and early *castra,* the camps of the fifth Macedonian and fourth Scythian legions, still could be found in the same parts of Macedonia as Justinian's home. Macedonia itself had been a Roman colony since 148 B.C.E., but the Illyrians mounted a resistance, and it took the army of Augustus, led by his general and successor Tiberius, to finally annex the region in 9 C.E. Even that proved inadequate, and subsequent campaigns under Domitian in 85, and Trajan in 101–106 were needed to pacify the territory.

By that time, Rome's policy was no longer to sow conquered ground with salt, but with retired servicemen. Illyricum was heavily colonized by ex-soldiers from the seventh legion who, while not ethnically Roman—most were from Spain, or Gaul, or Syria,* or even Africa—had been granted *ius italicum,* the right of Roman citizens, as

---

* The territory known in antiquity as Syria is somewhat vague, but is far larger than the modern country that bears the name, bounded by the Taurus Mountains in the north, the desert in the east, the Mediterranean in the west, and the Sinai peninsula in the south.

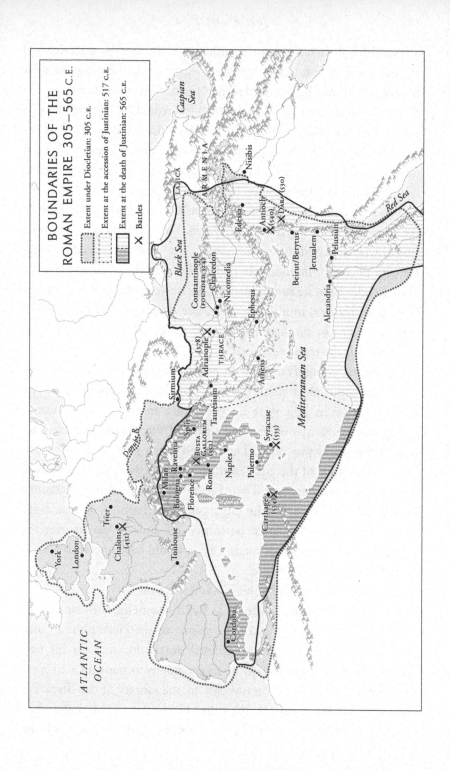

BOUNDARIES OF THE
ROMAN EMPIRE 305–565 C.E.

Extent under Diocletian: 305 C.E.
Extent at the accession of Justinian: 517 C.E.
Extent at the death of Justinian: 565 C.E.
X   Battles

ATLANTIC
OCEAN

Caspian
Sea

Red Sea

Black Sea

Mediterranean Sea

LAZICA
ARMENIA
THRACE

Nisibis
Edessa
Antioch (540)
X Dara (530)
Beirut/Berytus
Jerusalem
Pelusium
Alexandria
Ephesus
Athens
Constantinople (FOUNDED 324)
Chalcedon
Nicomedia
Adrianople (378) X
Sirmium
Danube R.
Split
Spalato
Taurésium
Busta Gallorum (552) X
Ravenna
Bologna
Milan
Florence
Rome
Naples
Palermo
Syracuse (555) X
Carthage (534) X
Córdoba
Toulouse
Chalons (451) X
Trier
London
York

a reward for their service to the emperor. That same long service had turned them all into Latin speakers, whatever the language of their homeland. Rome had rewarded her pensioners with homes in the Balkan highlands, and the new highlanders repaid loyalty with loyalty. Like Scots in Britain, or Virginians in the United States, Illyrians enlisted in the army in disproportionate numbers for generations, and it has truly been said that from the second century onward, the empire lived behind ramparts manned by an Illyrian army.

The road taken by Petrus Sabbatius had been paved by the boots of ten thousand of his Illyrian countrymen. One of them was his uncle, Justin, who departed sometime around 470 "after an unceasing struggle with poverty,"[2] intending to make his way in the world as a soldier. Two decades later, he summoned his twelve-year-old nephew, and the boy began his own month-long trip, one that began on the imperial highway that ran through the foothills of the Rhodope Mountains and ended at the empire's capital.

On arrival, young Petrus would surely have recognized it as the first city of the world, the richest, the most populous, and the most powerful. On its seven hills were an enormous, marble-paved forum, a stately and spectacular senate, and a palace from which a sizeable fraction of the world was ruled. The city housed a giant arena in which tens of thousands of the city's rowdier elements were kept docile by a diet of bread and circuses. Their bread was baked with the grain of Egypt, delivered daily to the city's gates and harbors, as was every other commodity that could be carried by mule or oxcart over the world's greatest network of roads. Petrus would have known it for the capital of the Roman Empire.

Just as he would have known it was not Rome.

The city was originally known as Byzas—sometimes Byzantium—for the Greek sailor who had led a colony to the Hellespont a thousand years before Constantine decided on the location for his new imperial capital. It was blessed by its geography, occupying a triangular promontory bordered on one side by the estuary of the Black Sea known as the Bosporus, on the other by the Sea of Marmara, into which that estuary drains. The site overlooked the inlet of the

Bosporus that would come to be known as the Golden Horn—"horn" for the shape that geology gave it, "golden" for the treasure flowing through it—straddling Europe and Asia, and dominating both. Less than three hundred miles from the barbarian tribes of the Danube, not much farther from the armies of Persia, the city was nonetheless virtually impregnable except by its landward side.

The city that greeted the boy Petrus Sabbatius was home to Greek and Latin speakers, Africans, Illyrians, Aramaic-speaking Jews and Syrians, Copts, Heruls, Gepids, Huns . . . "all seventy-two tongues known to man were represented in it,"[3] more than half a million people packed into a space roughly two-thirds the size of Manhattan . . . and, unlike Manhattan, it had no residence taller than a few stories in height, which meant that most of those people were found crowding the city's streets during daylight hours. The polyglot population amused itself by stereotyping newly arrived provincials: Armenians were lazy (a nun wrote that the unchurched Armenians were mean and ignorant, and that the Christian Armenians were even meaner and more ignorant), Cappadocians were rubes, Turks were deceitful, and Westerners— already starting to be called Latins—were impulsive and unsophisticated.[4]

Petrus, one of those unsophisticated Latins arriving from the west, had limited choices regarding his manner of entry into the city. Because the capital was surrounded by water on three sides, only two main roads led into it. The old road turned inland after passing the outlying town of Heraclea near Lake Bafa and entered the city proper just northwest of the Church of the Holy Apostles. The "new" road— built around 333—followed the coast until the point where it reached what came to be known as the "Golden Gate."[5] By either gate, the traveler entering from the west would find the last few miles of easting— the ones within the city, proceeding to the end of the peninsula, where the Bosphorus meets the Sea of Marmara—as filled with wonders as the entire journey leading up to it. The treasures of an entire world had been pulled into the imperial capital in wagons and on ships owned by Britons, Gauls, Africans, Persians, Indians, and Chinese, there to be displayed in houses of prayer, and those of commerce.

The squares, churches, and palaces of Constantinople are quite properly revered as the city's architectural jewels, but by far the most historically important construction in the city's history were the first things seen upon arrival: its walls. The walls sealed off the peninsula on which the city was built, turning it into a fortress that would stand, unbreached, for eleven centuries.

Along the *Mese,* the greatest of all the capital's boulevards (it remains so in modern Istanbul, whose main artery, the Divanyolu, follows the same path), tens of thousands of its residents could be seen doing the work of the city. Some of their activities would be familiar to any city dweller: buying, selling, seeking economic success and the advantages of status. Others would not, such as debating Christian doctrine. Gregory of Nyssa memorably described daily life thus: "If you ask about your change, the shopkeeper philosophizes to you about the Begotten and the Unbegotten; if you enquire about the price of a loaf, the reply is: 'The Father is greater and the Son inferior'; and if you say 'Is the bath ready?' the attendant affirms that the Son is of nothing."[6]

Though its public conversations might seem odd to modern ears— Gregory's bemused description suggests that they were regarded as exotic even then—its public economy would be familiarity itself. Constantinople, like Washington, D.C.,* was almost entirely a one-industry town whose industry was administration, and whose only products were laws. It was neither a center for industry nor—surprisingly, given its location at the knot in the bow tie connecting Europe with Asia— for commerce, at least not when compared with Alexandria.[7] The shopkeepers described by Gregory had, early in the city's history, segregated themselves into specific districts, with bronze smiths at the eastern end of the Mese, furriers near the center, and horse traders in the Amastrianum,[8] for the same reason that New York's diamond merchants congregate on Forty-seventh Street, or London's tailors on Savile Row: It was economically advantageous both for their cus-

---

* Also, as with Washington, D.C., the city's local government was administered by imperial officers.

tomers and their suppliers. Perfumers were granted the space closest to the Imperial Palace for a slightly different reason: The royals liked the aroma.[9] It was, no doubt, more pleasant than the city's most ubiquitous fragrance: the fermented fish sauce called *garam,* a taste for which was one of the many reminders of Constantinople's Roman ancestry. To the extent that other businesses flourished, they tended to be working either in the provision of luxury merchandise, such as silks, ivory, gold, and silver to the government functionaries, or in the maintenance of buildings, harbors, and aqueducts.

The workers in precious metals and vendors of silks and jewels had little difficulty in identifying their potential purchasers, the members of Constantinople's upper classes, particularly those of the Patrician Order, a formal honor dating back to the days of Rome's Republic. Patricians dressed in white, ankle-length tunics with narrow sleeves, edged with purple and belted with a red sash. Lacking either the advantageous proximity enjoyed by merchants of like products or a high society that rewarded residents with exclusive addresses, the great estates of the upper classes were built cheek by jowl with apartment buildings and even tenements; one consequence was that their mansions were, with few exceptions, inconspicuous from the street, reserving their luxurious appointments to their interiors, frequently built around open courtyards with bow windows on upper floors, permitting residents to see the street below.[10]

Though only a few hundred of the city's private residences could be called mansions, a very respectable percentage of the populace lived in single-family homes. Constantinople's fifth-century census counted 4,388 single-family dwellings, at a time when Rome had no more than 1,800;[11] given the size of families and household staffs, this probably represented between 5 and 10 percent of the entire population. What is, by ancient standards, such a high percentage, is largely due to a perpetual grant of a ration of bread—the *panes aedium*—which was not only given to anyone building a house in the city but passed along to the new owners with the sale of the house . . . a powerful incentive to both house builders and home buyers.[12] Even so, the street was home to a large number of Constantinople's residents, possibly even a majority, who lacked any sort of permanent lodging, and depended upon homeless shelters run by the city's monasteries and churches.

They also depended on the dole. Like Rome, where grain, and later bread, had been given away to citizens since the time of the Republic, Constantinople offered a free bread ration to a substantial number of its citizens, some by hereditary right, as with homeowners and some scholars, but many more on the basis of need. The policy was driven less by charity than by a well-founded fear of insurrection. Both Rome and Constantinople were chronically at risk of riots caused by food shortages, usually though not always the result of speculation. To prevent disturbances, the civic authorities, which controlled retail prices for pork, beef, and wine (though not for fish, a poor choice for hoarding) directed the one hundred public bakeries of Constantinople to provide two pounds of free bread to each of eighty thousand residents daily.[13]

The newly arrived provincial was likely to be more impressed by the capital's monuments than its bakeries. Following the Mese on its ruler-straight path upward through the hills of the city, the traveler would pass the artifacts that represented the final victory of Christianity over paganism. Underneath the column erected at the center of the city's central forum were not merely fragments of the True Cross, and the nails that held the Savior to it, but Aeneas's icon of Athena, and the palladium of Troy.[14] The path continued, as did Petrus Sabbatius himself, to the eastern rim, which was not only the original summit of the city but also of the entire known world. At its center was the great square, the Augustaion, an inside-out ziggurat composed of rows of seats rising in a square helix, each step slightly higher than the one next to it. To its east was the Senate House, fronted by six enormous columns, two set back slightly, the other four forward, collectively supporting a vaulted marble entry decorated with dozens of statues; to its north were the equally gigantic Baths of Zeuxippus. Baths were ubiquitous in the city; by the fourth century, Constantinople housed 8 public and 153 private ones, fed by aqueducts that brought water from the mountains of Thrace and stored it in underground reservoirs until needed.[15] Within a single square mile that constituted, both psychically and geographically, the tip of the last finger of the long arm of what would one day call itself Europe, stood both the Great Palace from which the emperor ruled, and the Church of Holy Wisdom—the Hagia Sophia—in which he accepted the rule of God, who had, not at all coincidentally, first revealed Himself in the east.

When Petrus left Tauresium, he had traveled east, away from the empire's past, and toward its future. His choice of destination had effectively been made for him by a series of unrelated events. They include the end of the empire's most stable series of rulers, the five Good Emperors, beginning with Nerva and ending with the death of Marcus Aurelius; the so-called Crisis of the Third Century that followed, which featured a fifty-year period during which twenty-six emperors reigned* and twelve were murdered, mostly by their own Praetorian guardsmen; and the rescue of the empire from its seeming free fall by yet another series of rulers, the Soldier-Emperors, most from the Balkans, who ended the Third-Century Crisis with Rome ruling a territory as vast as ever, but of a vastly different character. One of the last of them, Aurelian, who not only reestablished the empire's borders, but built a defensive wall around the city of Rome, fully deserved the title with which the Roman Senate honored him, *Restitutor Orbis*: "restorer of the world."

It was left to another son of the Balkans to remake it.

Had Justinian departed his Balkan home two hundred years earlier, before Constantinople could exert her irresistible pull, he would have been far more likely to travel west rather than east, and, only a week or two after leaving Tauresium, would have arrived on the Adriatic coast. Heading north for another week would have brought him to a rather remarkable house, constructed on a sloping headland overlooking the middle of three inlets on the Dalmatian coast, on a semicircular bay facing south. Rugged mountains nearly five thousand feet high

---

* Twenty-six or more. Not included are a number of usurpers, nor those who declared themselves "emperor" of Gaul or Britain. Nonetheless, the number is dizzying; in 238 alone, five different emperors served, one for less than a month.

dominate the view to the north, while to the south, several small islands break the surface of the blue Adriatic. In between lies a verdant
strip, cultivated with olive and fig trees and grape vines.[16] The climate
is temperate. All in all, it was a most pleasant spot for a typical retirement home.

There was, however, nothing at all typical about this particular
home, beginning with its size. Built to stand comparison with the
palaces of Egypt and Babylon, it was enormous, the main building
alone nearly ten acres in extent, with walls seven hundred feet long
and seventy feet high. The house's occupant was likewise atypical, a
former soldier who had been, for twenty years, the most powerful
man in the known world. Still more remarkable was the fact that the
palace was occupied at all. Roman emperors died in battle, or of disease, or were assassinated. They didn't retire.

Most remarkable of all, when the retiree—the first Roman since
Augustus to found a truly new empire—died, the event was so *unre-
markable* that no one can say with certainty the year it happened. His
parents, likely freed slaves, had named him Docles when he was born,
and he died using the same name, despite some rather dramatic
changes in circumstances. During the intervening sixty or seventy
years, he accumulated names with the same avidity that he earned honors, finally taking the throne as Gaius Aurelius Valerius Diocletianus—
the Emperor Diocletian.

Like so many of his Illyrian countrymen—he was born in the same
village where he would build his palace, in Salona (the city Split in
modern Croatia)—Diocletian rose to prominence with a sword in his
hand, serving as a soldier, later a general, in a war against the Persians.
In 285, after what had become almost the normal pattern of transfer-
ring political authority in late third-century Rome—emperor killed
on campaign, heir found dead of mysterious causes, adoptive father
accused of murder and executed, emperor's eldest son defeated in
battle—Diocletian succeeded to the empire of Augustus. Almost immediately, he set about transforming it forever. In 286 Diocletian recruited a coemperor, a fellow Illyrian named Maximian, and formally
divided responsibility for imperial defense along an east-west axis.

Equally important, neither emperor made his capital in Rome, with Maximian established at Milan and Diocletian ruling from Nicomedia, present-day Izmit in Turkey. Seven years later, Diocletian expanded the administrative responsibility for the forty-four provinces of the empire into a tetrarchy, adding two more corulers: Galerius and Constantius, with the former based in the Rhineland city of Trier, and the latter in Sirmium (today in Serbia). The second pair of "emperors"—the two "Caesars"—were adopted by the first two (each an "Augustus"), and connections were further reinforced by intermarriage; the four tetrarchs were subsequently called the *quattuor principes mundi*: the four princes of the world.

The life of Diocletian has served for centuries as an object lesson in the value of rational intelligence over impulsive genius. Gibbon's report card still rings true:

> The valour of Diocletian was never found inadequate to his duty, or to the occasion, but he appears not to have possessed the daring and generous spirit of a hero, who courts danger and fame, disdains artifice, and boldly challenges the allegiance of his equals. His abilities were useful, rather than splendid—a vigorous mind improved by the experience and study of mankind; a judicious mixture of liberality and economy, of mildness and rigour.[17]

The civil accomplishments of Diocletian and his other tetrarchs are considerable. Their domestic reforms include dramatic changes in the tax laws, restoration of the currency—the *aureus* was standardized at one-sixtieth of a pound* of gold, and a new silver coin, the *argenteus,* replaced the devalued *denarius*—and writing a hundred new laws . . . without benefit of the Roman Senate, which had lost the last vestige of its lawmaking powers under the emperor. The tetrarchy professionalized the civil service, and built entire cities from the Danube to the Nile Valley, essentially turning the entire empire into a set of urban administrative zones.

They also transformed imperial military strategy. Despite the many

---

* A Roman pound, about twelve ounces.

wars of the third century, the army in place when Diocletian took the throne was similar in organization to that of seventy years previous, though considerably larger. In truth, it hadn't really changed since the time of Augustus, whose army consisted of 150,000 men divided into twenty-five to thirty legions, each one comprising approximately 5,500 infantry troopers armed with short sword, javelin, shield, and armor. By Diocletian's time, the army fielded perhaps sixty similarly sized legions,[18] between 350,000 and 400,000 men: One contemporary historian gives the figure of 390,000,[19] while another estimates 645,000 by mid-fourth century.[20]

But Augustus's army was an army of conquest. Diocletian* reconfigured it into an army of defense—or, to be precise, two armies. The first, the *limitanei,* occupied a buffer zone along rivers like the Danube and the Rhine, dotted with strongholds and garrisoned with border troops. The second, a "corps" of which was generally based near one of the tetrarchs, was a mobile field army, or *comitatensis,* formally attached to the emperor's household—the *comitatus*—for the duration of a campaign.[21] The armies worked in tandem, one to hold cross-border invaders in place long enough for the other to crush them.

Recent experience, however, had taught that a strategy that placed armies permanently in the field (three *comitatenses,* of about thirty thousand men each, were based in Gaul, Illyricum, and the east) also required one of the tetrarchs nearby as a deterrent to overambitious generals. As a result, while Rome remained the home of the ever-more impotent senate, the real power migrated to cities like Milan, Trier, and Nicomedia. Added to its geographic disadvantage was Diocletian's personal distaste for the Eternal City: The emperor waited until the celebration of the twentieth anniversary of his reign—he was the first emperor since Marcus Aurelius who lasted long enough to enjoy the opportunity—before visiting Rome for the first time, and he left after only six weeks. In Gibbon's elegant and accurate phrasing, "The dislike expressed by Diocletian towards Rome and Roman freedom

---

* Though, to be sure, the transformation started before Diocletian, and continued after him. Thirty years ago, Edward Luttwak started the debate whether the grand strategy was planned or opportunistic, a debate that continues to this day.

was not the effect of momentary caprice, but the result of the most artful policy."[22]

The policy was intended not merely to relocate authority from the empire's historical core closer to its periphery, but to formalize a peaceful imperial succession—an understandable priority after a century in which the most frequent cause of death among Rome's emperors was assassination. Seventeen centuries after the Third-Century Crisis, the *de minimis* characteristic of stable government remains the ability of a nation's leader to transfer executive power peacefully ... and, as scarcely needs repeating, such a peaceful transfer of power remains out of the reach of billions of people to this day. So, to all of Diocletian's achievements, add this: Less than two years after his twentieth anniversary celebration, on May 1, 305, Diocletian assembled his court in Nicomedia and announced his abdication in favor of his "junior" emperor, Galerius; on the same day, in Milan, the other Augustus, Maximian, also abdicated, retiring to Campania. Rome looked to be in permanent decline as a political center, and the four-headed structure that replaced it seemed robust indeed.

———

In the end Diocletian's plan for the peaceful transfer of imperial power depended on successors who had a stronger sense of duty to the state than an urge to start a dynasty. As such, it was problematic from the start. The two Caesars who were made Augusti upon the abdication of Diocletian and Maximian—Galerius and Constantius—were themselves replaced as Caesars by Maximin Daia (junior to Constantius) and Severus (junior to Galerius). Constantius's son, who had previously been tapped as successor material himself, was cut out of the imperial succession, and held against his will at the eastern court as hostage against his father's good behavior ... hostage, that is, until he escaped, riding the breadth of the Mediterranean to join his father, the newly titled Augustus Constantius in southern France. The son was called Constantine.

As had become almost compulsory for potential emperors, Flavius Valerius Constantinus was Balkan-born, in Naissus (Niŝ) sometime between the late 270s and 280s and—equally compulsory—had distinguished himself both as a soldier and as a courtier while serving in

Diocletian's court at Nicomedia. Gibbon, though grudging in his assessment of Diocletian, positively gushes over Constantine:*

> The person, as well as the mind, of Constantine, had been enriched by nature with her choicest endowments. His stature was lofty, his countenance majestic, his deportment graceful . . . the disadvantage of an illiterate education had not prevented him from forming a just estimate of the value of learning; and the arts and sciences derived encouragement from the munificent protection of Constantine. In the dispatch of business his diligence was indefatigable, and the active powers of his mind were almost continually exercised in reading, writing, or meditating.[23]

By the time his father died while campaigning in York on January 25, 306, Constantine had impressed the British legions sufficiently that they almost immediately acclaimed him as the new Augustus, his father's successor. For six years, he regarded himself as such, even though Galerius, the "other" Augustus in Nicomedia, would only agree to regard him as a Caesar. To further muddle things, Diocletian's former coemperor, Maximian, had renounced his reluctant abdication—it seems in retrospect that the only person who believed in Diocletian's plan was Diocletian himself—and with his son, Maxentius, asserted imperial authority over the Italian peninsula. When Galerius died in 311, the muddle turned bloody; Maxentius publicly accused Constantine of rebellion, in effect declaring war on him. The new emperor in the east, Licinius, was likewise warring with *his* junior emperor, Maximin Daia. As a result, though he always dated his reign from the death of his father, Constantine's authority was distinctly limited until the end of the civil wars that put paid to Diocletian's tetrarchy.† The

---

* The pre-Christian Constantine, anyway. Relying on pagan historians like Eutropius, and obedient to his own thesis that Christianity was the primary cause of the "decline and fall," Gibbon regards the second half of Constantine's life as largely subtracting from the achievements of the first.

† When Diocletian was importuned by Maximian to return and end the civil wars, the former emperor famously replied that if Maximian could see the cabbages he had cultivated with his own hands, he would never ask that he "relinquish the enjoyment of happiness for the pursuit of power."

most famous battle of those wars earned its notoriety not merely because of its military significance, but as a key moment in the history of religion.

Constantine invaded Italy in the spring of 312, and by fall, after winning victories in battles fought outside Turin and Verona, had arrived outside Rome. Troops loyal to Maxentius, notably the Praetorian Guard, which still formed an elite among Rome's armies, and which had made and unmade emperors for a century, made their stand at Saxa Rubra, nine miles from Rome, where, on October 28, 312, they faced off against Constantine's outnumbered but far more experienced troops. By the time the battle had ended, Maxentius's army had been outflanked and forced over the Milvian Bridge that led into Rome, and the bodies of the Praetorians "covered the same ground which had been occupied by their ranks."[24] Maxentius was dead, and Constantine became the undisputed ruler of Rome's western half. But the real import of the battle of Milvian Bridge is due to an event that occurred the night—or the day—before.

The specifics of Constantine's famous vision are generally reported in one of two ways. In the earlier version, as recorded by the chronicler Lactantius, he dreamed that he was ordered to paint two Greek letters, superimposed on each other, on the shields of his troops;* in the version set down years later by the historian Eusebius, a cross appeared in the sky on the day before the battle accompanied by the legend *in hoc signe vinces*: "by this sign you shall conquer." Later speculation about Constantine's motives notwithstanding—some have suggested that what the emperor actually saw was a conjunction of the planets Mars, Saturn, Jupiter, and Venus in an astrologically troubling part of the night sky, a conjunction that might have frightened Constantine's army had he not found an inspirational angle in it—the results were world-historical in their impact. An emperor at the height of his power believed himself in debt to the Christian god.

Less than a year later, Constantine met with his opposite number, Licinius, in Milan, and the two emperors not only agreed to divide

---

* The letters, X and P, the first two letters of "Christ" in Greek, had long been a Christian monogram.

the empire between them, but jointly issued an edict granting official toleration to Christianity. The first accord collapsed in 324, when war between Licinius and Constantine ended with the defeat of the former and the accession of the latter to sole rule over the empire. The second agreement, however, would cement the bond between empire and church for a thousand years. For Constantine agreed not merely to tolerate Christianity, but to embrace it. Over the next thirteen years, the new ruler left his mark throughout the empire, and more often than not, the mark he left was in the shape of the cross.

———

At the moment when Licinius and Constantine agreed to the Edict of Milan, the wealth of the east, the wars of the Soldier-Emperors, and the policies of Diocletian had, at least temporarily, separated Rome from its empire. The city was declining in both population and political clout, and suffering from the relative poverty of the western provinces—the self-sufficient *latifundia*, or plantations, that composed the great fortunes of the west were unable to produce wealth comparable to the trading economy of the east; though it might be a simplification to see a resemblance between the third century's western and eastern territories and early nineteenth-century Massachusetts and Virginia, it is not an egregious one. Rome was not yet a complete ruin, however, and retained some powerful claims to political authority, none more important than the historically dubious but widely accepted belief that the Apostle Peter had been martyred there, sometime during the reign of Nero. As the final resting place for the church's first leader, the preacher at the first Pentecost, the man whom Jesus called the rock—*cephas* in Aramaic; *petrus*, hence "Peter"—on whom the church would be built, Rome could, and did, claim pride of place among all Christian bishoprics.

It was a convincing choice for the capital of the first Christian empire, and in the early years of his reign, the first Christian emperor seemed to agree. Constantine donated the Lateran district to Bishop Melchiades,* and within the Lateran built a new cathedral. Rome's

* Retrospectively Pope Melchiades. For centuries, *papa* was a term applied to many bishops, usually as a sign of a personal relationship.

claim was not, however, preemptive; many more Christians lived in Alexandria, Antioch, and Jerusalem than in Rome. Diocletian and his coemperors had ruled from cities like Nicomedia and Milan. Any of them might serve as Constantine's capital. No doubt some of the factors that conspired in Constantine's choice were traditional ones: political advantage, military defensibility, geography. In the end, the decisive one grew out the most passionate religious dispute of late antiquity.

In 306, as Constantine was accepting the homage of the Roman legions in Britain, Arius, an Alexandrian priest with a powerful combination of great intellect and imposing presence, "tall and thin, with a melancholy aspect,"[25] broke with his nominal superior, the bishop Alexander, in arguing that Christ, while an intermediary of God (and, in some way, a reflection of divinity) could not himself *be* divine. Arians subsequently appeared in both diluted form ("Christ is a supernatural being, God's first creation, but not his equal") and unfiltered ("Christ is not divine at all"), but always saw themselves as warriors in a battle against polytheism.

The new emperor was neither educated for, nor particularly interested in, theological hairsplitting. Moreover, he was on record as believing that the Arian position was a trivial heresy—if such does not seem self-contradictory—and easily resolved, writing (in a letter to Alexander and Arius) that their "cause seems quite trifling, and unworthy of such fierce contests . . . problems whose only use is to sharpen men's wits."[26] To say that he misunderstood the passion with which it was held is to seriously understate the case.

The conclusions of Arius—not only that Jesus, created by God, could not share his creator's timeless character or substance, but that the Holy Spirit, likewise created from God's highest created being, occupied an even lower rung on the ladder that led from God to man— hit the eastern empire like an earthquake . . . and like an earthquake, was followed by equally powerful aftershocks. In 320 Arius was excommunicated by a group of nearly one hundred bishops assembled by Alexander. But his ideas continued to spread throughout the empire. Constantine had to intervene in the dispute tearing at the heart of his world, and he chose to do so by calling the first Ecumenical Council in church history.

On May 20, 325, more than three hundred bishops, largely from the empire's eastern provinces, convened in the imperial palace of the town of Nicaea, close by Diocletian's capital of Nicomedia. Opening the council, watching over its deliberations, and demonstrating a mastery of politics as sophisticated as his philosophical inquiries were simple, was the emperor himself. It was Constantine—robed in purple, glittering with jewels, he made certain to ask permission of the bishops before taking his seat . . . but seated himself first each day— who determined, as the questioning of Arius took on an ever more prosecutorial tone, that it was more important that the conference reach consensus than final truth. And it was Constantine, consistent in his belief that anything that weakened either empire or Church weakened both, who contributed a theological concept that split the difference between the Arian belief that Jesus's substance could not be divine and the orthodox conviction that it must be. The emperor's inspiration was to replace the Arian concept of *homoiousia*, or *similar* essence, with *homoousia* or *same* essence: consubstantiality; it has been well said that the Christian world had been torn apart, and then repaired, by a single letter, the Greek iota. Even more important, he encouraged the vaguest possible definition of the concept, in order that the final formula of the council, the original Nicene Creed,* would have the largest number of signatures of the bishops present. He got his wish: When the council ended on June 19, only two refused to sign, and joined Arius himself as excommunicants.

The Council of Nicaea proved such a coup for Constantine that he used its conclusion as a fitting opening for the celebration of the twentieth anniversary of his reign. Afterward, in January 326, the emperor and his family departed for Rome, there to continue the festivities. It was to prove an unhappy journey: Unhappy for the emperor's family, three of whom he had killed en route for rea-

---

* Somewhat different from the one still in use today, which was codified in 362.

sons that remain mysterious;* unhappy for the emperor, who took offense when requested by Rome's aristocrats to take part in a pagan ceremony; and unhappy for the future of Rome, for when the emperor departed the city—early, like Diocletian a generation before—he did so determined to build a New Rome that would surpass the original.

Even before the anniversary celebration in Nicaea, "Constantine the celebrated emperor . . . renewed the first wall of the city of Byzas and joined them to the ancient wall of the city and named it Constantinople."[27] Now, in 328, he started rebuilding his new city, his *Nova Roma,* in earnest: He "built a Forum which was large and exceedingly fine, and he set in the middle a great porphyry column of Theban stone . . . and placed [the Palladium] in the Forum built by him, beneath the column of his monument. . . . Nearby he also built a basilica with an apse, and set outside great columns and statues; this he named the Senate."[28]

The capital of a Christian empire was an ideal setting for displaying the spoils of the pagan peoples conquered or supplanted by the Church, including the obelisks of Egyptian pharaohs, the Serpent Column from the Temple of Apollo in Delphi, and the famous horses, supposedly sculpted by Lysippos, the great fourth-century B.C.E. sculptor (the only sculptor for whom Alexander would sit), and brought by Nero to Rome for display in his Domus Aurea, or Golden Palace.† But the city would house not just displays of Christianity's victory, but evidence of its birth.

A year earlier, in 327, the emperor's mother, Helena, had visited Jerusalem, earning a place in history as the first pilgrim to the Holy Land.[29] Not surprisingly, she found treasures everywhere, one of which

---

* The dead included the emperor's wife, Fausta, and first-born son, Crispus. Zosimus, a historian writing a century after the events, suggests that the two were entangled sexually.

† Those horses are some of the most traveled statuary in history, captured by crusaders in 1204, who brought them back to San Marco, where they stayed until Napoleon Bonaparte took them as spoils to Paris in 1797. Upon his defeat in 1815, they were returned to Venice, where they remain to this day.

was the most important site in Christian history, the other its most important relic. Over the first, her son built the Church of the Holy Sepulchre. The other—the True Cross itself—was placed by the emperor inside the dome that sat atop four columns at the *milion*, the square at the center of his new city. Radiating from the *milion*, Constantine's great boulevard, the Mese, ran west to his forum, dominated by a great statue of the emperor himself, and east to the luxurious new senate, the emperor's palace, and the church of St. Eirene, where, on May 11, 330, the new capital of the newly Christian Roman Empire was formally dedicated.

———

The city of Constantine was only 140 years old, therefore, when Petrus's uncle Justin arrived, enlisted in the *excubitores,* an elite regiment of the Roman army, and started a long climb up the officer ranks. Along the way he married a former slave girl named Lupicina, and settled in the imperial capital, behaving like a million other provincials-made-good before and after: He sent money home. Nor did he stop there, summoning half a dozen Illyrian relatives to join him in Constantinople, one of whom was his nephew. Having escaped the bone-grinding poverty of his Illyrian ancestors, Justin was eager to share his good fortune with his family. But while he had made his fortune with his sword, he had something different in mind for his favorite protégé. While Justin soldiered, Petrus studied.

The political leaders of great imperial cultures are educated to rule, and in sixth-century Constantinople, the career-making importance of education was very high indeed. The Soviet Union raised its children on Marxism-Leninism; Great Britain on some combination of Eton, Oxbridge, and Kipling. The sixth-century Roman version was a pedagogical tradition that would one day be called Neoplatonism.

Neoplatonism—Justinian's teachers were pious, but sophisticated, Christians, and they did not call it that; if they called it anything, the word they used was Platonism—offered its acolytes a fully formed and complete view of the world. Derived from the almost willfully obscure work of Plotinus, a third-century Egyptian philosopher, and Christianized by the most brilliant bishop of the day, Origen of Alexandria, Neoplatonism taught the existence of a transcendent realm, in which the seeming conflicts of lower domains were reconciled. The belief in a purer, simpler, more profound, and—most significantly—*unconflicted* world is the most durable idea of Neoplatonic education; not for nothing do Neoplatonists call the highest rung on the ladder of reality, the "One."

So it was that, during the last years of the fifth century until 505 or so, while his uncle practiced the disciplines of a soldier and commander of soldiers, young Petrus Sabbatius spent his time in the company of Origen's heirs, absorbing transcendence along with Latin declensions. His reward for such diligence was less the mastery of prose—his Latin style has been criticized by moderns and contemporaries alike as vulgar, simultaneously pretentious and colloquial[30]—than the skills of a philosopher, along with what can only be called a singleminded faith in singlemindedness. Many years later, when the future emperor had reestablished imperial rule in Italy, and his courtiers argued in favor of a fully Christian educational system in the peninsula, he proved faithful to his original teachers, famously refusing them "so that young men trained in liberal arts might continue to prosper in my empire."[31]

Justinian may well have been loyal to the idea of a single law and religion for a single empire when he arrived in the capital; Rome had commanded such loyalty for centuries before Christ, to say nothing of Neoplatonic Christianity. It is certain that, by the time Justinian entered the historical record as a young adult, he regarded anything that threatened the unity of the empire as an affront against reason and faith alike, and the most persuasive explanation was his dual inheritance: that which had been rendered both unto Caesar and to God. On circumstantial evidence alone, he seems more than a little like Alexander, learning the universal superiority of Greek thought from Aristotle. Even more like Alexander, Petrus, a Latin-speaking immigrant in a Greek-speaking world, was a child of the provinces, with a convert's enthusiasm for his adopted world.

It is therefore noteworthy that by the time he completed his education, the single empire to which the Illyrian peasant had sworn loyalty had not been a reality for two hundred years, and had not even ruled Rome itself for half a century.

CHAPTER TWO

"We Do Not Love Anything Uncivilized"

337–518

I N THE CITY of Nicomedia, seven years after the dedication of the
new imperial capital, a Christian neophyte in his sixties died, and
was buried in the white robes that marked his acceptance into the
church. Deathbed conversions have never been especially unusual and
were highly respected among serious fourth-century Christians, who
frequently postponed baptism until past the time when they could
commit any more sins. In this, Rome's first Christian emperor
planned well: Constantine first accepted the sacraments from Bishop
Eusebius only weeks prior to his death, having committed his last
great sin well before.

Having long exhausted all the possibilities of avarice, envy, and
even anger, the emperor nonetheless remained vulnerable to an auto-
crat's pride, and made the politically disastrous decision to provide for
his succession by naming five junior emperors, five "Caesars": his
three sons and two of his nephews. The results were predictable, recalling the bloodiest years of the Claudians—or possibly a Shake-
spearean tragedy. First, the emperor's half-brother Delmatius, and
then Delmatius's sons were assassinated by one or more of their rival
cousins. Constantine's other half-brother (through his father's second
wife), Julius Constantius, was next. Remaining were Constantine's
three surviving sons, with the confusingly similar names Constantine,
Constantius, and Constans. Constans defeated and killed Constantine in 340, was killed in turn in 350, and Constantius became sole
ruler in 353.

But despite the calamitous results within his own family, each of
Constantine's world-historic decisions survived his death. Had Rome's

first Christian emperor been granted the ability to see two hundred years into the future, he would certainly have been cheered to see that Constantinople remained the imperial capital; it would, in fact, remain so for eleven centuries, until the cannons of Sultan Mehmet conquered the fortress city in 1453. To adherents of what the modern world calls Orthodox Christianity, it remains a capital to this day. The empire's status, however, depended upon the extent of its dominions and had Constantine been able to see the future of his western domains, he would have seen the homeland of Rome's empire ruled by a people who were not even the nominal heirs of Augustus.

Almost from the day of Constantinople's founding, the boundaries of that empire started to appear problematic. The emperors that followed Constantine, even those presiding over steadily shrinking borders, would remain firm in their belief in the eternal character of the empire, and its *de jure* sovereignty over all the civilized peoples of the Mediterranean world.

Their ability to rule the uncivilized, however, was coming to an end.

The name Goth (in the form *Guti* or *Gutones*) started to appear in chronicles during the early first century, and is mentioned by Ptolemy in his geography around the year 150. The Goths are mentioned as one of the peoples defeated in battle by the first of the Persian Sassanid kings in 245; and in the year 269, the Roman Emperor Claudius II took the name Gothicus commemorating his victory at Naissus the preceding year.[1] It is certain that for forty years, beginning in 238, Goths ravaged Roman settlements on both sides of the Danube, inevitably drawing attention from, and then retribution by, the empire.

What their sixth-century contemporaries knew of the Goths' early history was largely due to their first chronicler, Jordanes, who set down his *De origine actibusque Getarum* (usually translated as the "origins and deeds of the Goths" and referred to in shorthand as the *Getica*) in 551. The *Getica*, Jordanes's abridgment of a much longer, and now lost, work by Cassiodorus, a retainer to the Goths' greatest king, is more a national epic than a reliable chronicle, a better guide to what the Goths believed about past events than to the events themselves. Reading it, one would learn that the Goths' ancestral home was in southern Sweden; that their legendary King Berig led them across the Baltic in three ships, where they defeated the tribes then occupying what is now Lithuania and settled. The *Getica* tells the story of their adventure-filled migration—the legend includes encounters with Darius the Great, Alexander, and Julius Caesar—south to the Black Sea, where they enter the historical record proper.

The Goths owned a large share of the Third-Century Crisis:

> While [the empire] was given over to luxurious living of every sort, Respa, Veduc, and Thuruar, leaders of the Goths, took ship and sailed across the strait of the Hellespont to Asia. There they laid waste many populous cities and set fire to the renowned temple of Diana at Ephesus, which, as we said before, the Amazons built. Being driven from the neighborhood of Bithynia they destroyed Chalcedon, which Cornelius Avitus afterward restored to some extent. Yet even today, though it is happily situated near the royal city [Constantinople] it still shows some traces of its ruin as witness to posterity. . . . After the Goths had thus devastated Asia, Thrace next felt their ferocity.[2]

Aurelian had faced Gothic ferocity as an imperial general many times before he became emperor himself. The experience he had acquired in the field evidently persuaded him that the best solution to the problem of the Goths was to separate from them, along the Danube River, so when, as emperor, he reestablished the borders of the empire, he gave up the province of Dacia, on the northern bank of the Danube, as a reasonable price for peace.

And, for nearly a century, peace is precisely what his decision bought. Goths even served as allies to Roman emperors from the time an auxiliary unit served in third-century Arabia[3] and sometimes found themselves caught between them, as when they supported Licinius against Constantine in the 320s. Despite choosing the losing side, the Goths were quick to accept the religious faith of the winner. Their own Christianization followed only decades after Constantine's at the hands of Ulfilas, the first bishop to the Gothic Christians.

By a happenstance of history, the Gothic conversion was a problematic one. Ulfilas was converted himself, and named bishop in 341, while in Constantinople as part of an embassy from the then pagan western Goths. At that time, the metropolitan bishop of Constantinople—the man who performed Ulfilas's consecration—was a follower of Arianism. As a result, the people to whom Ulfilas returned, and where he preached for the next four decades, became Arian Christians as well, a matter of little consequence during those years when the empire was itself ruled by Arians, but a provocation once Arianism was declared heretical.

Had he done nothing but bring the Goths to Christianity, Ulfilas would have earned a large entry in any history of Europe. For those with a high regard for the power of the written word, his stature is even higher. Not only did Ulfilas create the first alphabet of the Gothic language, and translate the Bible into that language—the first translation of the Bible into any language of the family that would be called Germanic—but by doing so, he owns a piece of the responsibility for the romantic perspective modern Germans developed about their own history.

Generations of European ethnographers have mapped the "peoples of Europe" in late antiquity to those occupying the same territories on a modern atlas, but in some respects their work is the scholarly equivalent of genealogical social climbing, building family trees in search of an appropriately glamorous ancestry. Those pedigrees have been so successfully publicized, particularly by scholars writing in German, that most reference works describe the Goths as a Germanic tribe to this day, despite the fact that, in the words of their great modern historian Herwig Wolfram, "a history of the Goths is not part of the history of the German people."[4] And, if one defines history as a sort of genetic inheritance, he is correct: One's history is no more controllable than is the great-grandparent who bequeathed her descendants the gene for red hair.

However, history is not ancestry. The creation of Europe—specifically, the creation of that part of Europe we call Germany—was a work of the human mind as much as geography, and as such was unrestricted by the law requiring cause to precede effect. In that sense, the Goths *are* a part of German history, in the same way that Uncas, the eponymous hero of James Fenimore Cooper's *Last of the Mohicans*, is part of the history of New York, or Arthur the history of Britain. The beliefs of a people are no small part of their identity.

After Ulfilas, the core of Gothic identity was a shared language and religion; but while Ulfilas might have been bishop to all the Christian Goths, his flock was a diverse one, and the names by which its pieces were known were inconsistently applied until the sixth century. By then, the Gothic peoples living between the Danube and Dniester were known as the *Tervingi* or *Vesi* (hence Visigoths); those farther east, the *Greutungi*, though others knew them as *Ostrogothi*.⁵ Cassiodorus's distinctions (as filtered through Jordanes) in which the Ostrogoths appear as the "real" Goths, ascendant over their Visigoth cousins, reflect nothing so much as the perfectly understandable desire of a courtier trying to concoct a historical justification for the authority of his king's family. The names of the ruling families add yet another layer of confusion to Gothic history, since they are frequently used as synecdoches for the Gothic peoples themselves. *Amali* was not merely the family name of the hereditary rulers of the Ostrogoths, but was sometimes used as a synonym for them; similarly, the *Balthi* dynasty was the first family of the "other" Goths, whether known as Tervingi, Vesi, or Visigoths.

Constantinople was less concerned with the lines distinguishing Gothic peoples from one another than with the line that separated them from the empire. Tervingi or Vesi, Greutungi or Ostrogothi, Amali or Balthi, the Danube border held firm until finally breached by the ravages of a people even more ferocious than the Goths.

———

Imagine an avalanche. Imagine that the slope down which the avalanche falls is a five-hundred-mile-wide ribbon of grassland, stretching from China's river valleys and the Altai Mountains of southern Siberia to the northern shores of the Black Sea. Imagine standing at the terminus of the slope, watching the avalanche, which has been uprooting trees, smashing buildings, and destroying everything in its path for every one of three thousand miles, and you may be able to approximate the impact of the arrival of the horsemen of

the Eurasian steppe on the more settled peoples of the west. The journey was first taken in the second century B.C.E. by the nomadic
mounted archers known as Scythians, whose precise origin remains
unknown, since the practice for centuries was to call every Central
Asian horse archer a Scythian. They were followed by waves of other
nomads, some of whom have bequeathed their names to modern
Central Asian republics like Kazakhstan and Uzbekistan, but the most
destructive of the steppe peoples to appear in the historical record before Genghis Khan's Mongols were the Huns.* They presented a terrifying aspect to the fourth-century soldier-turned-historian Ammianus
Marcellinus, who wrote of them,

> From the moment of birth they make deep gashes in their children's
> cheeks, so that when in due course hair appears its growth is checked
> by the wrinkled scars; as they grow older this gives them the unlovely
> appearance of beardless eunuchs. They have squat bodies, strong
> limbs, and thick necks, and are so prodigiously ugly and bent that they
> might be two-legged animals, or the figures crudely carved from
> stumps which are seen on the parapets of bridges.[6]

Bad as they looked, they behaved worse, "uprooting and destroying
everything in [their] path . . . bringing sad calamity from the east . . .
like a whirlwind descending from high mountains."[7] The whirlwind
first descended upon the Alans, another steppe people occupying the
high Iranian plateau, who were defeated by the Huns at the battle of
the Tanais River in 371.[8]

The first animals that escape from a forest fire do so because they
see flames, or smell smoke. Other animals see them, and flee in the
same direction. The Hun firestorm had the same effect. The first
Goths to encounter them, the Ostrogoths, were conquered in the late
370s because of their exposed easterly position. Their fate convinced
the leaders of the other Goth tribes that they needed to put some

---

* Hun origins remain the object of a never-ending debate, with some historians arguing, partly
on the basis of similar-sounding names, that they are the same as the Hsiung-nu [alt.
Xiongnu], a nomadic tribe that ravaged Han China in the second century B.C.E.

natural boundary between themselves and the Huns. The southern shore of the Danube looked very nearly perfect.

By then, Constantine's brief dynasty had ended, and two brothers named Valentinian and Valens, both former soldiers and Arian Christians, had divided the empire between them, Valentinian in the west, and Valens in the east.

Valens had been emperor for twelve years when, in 376, the Goths, facing not merely the Huns, but tribes like the Alans who were themselves being pressed by the invaders from the steppes, asked for permission to settle in Thrace, primarily because "it is separated by the broad stream of the Danube from the regions exposed to the thunderbolts of the alien Mars."[9] According to the chronicler Zosimus, Valens agreed, on the condition that the Goths surrender their arms. None of the rich descriptions of Hun ferocity is as eloquent as the simple fact that the Goths agreed.

During the spring of 376, the Roman Empire supervised the crossing of the Danube by approximately 75,000 Goths[10] "on boats and rafts and canoes made from hollowed tree-trunks"[11]—a fourth-century version of the evacuation of Dunkirk. Unfortunately, the resourcefulness that the Romans exhibited in transporting the population of a small twenty-first-century city across the Danube failed them when it came time to feed their new charges. The Goths began to starve, though whether cupidity or an overstrained supply system was to blame is unknown. Likely it didn't matter to those Goths who were forced to sell their possessions, even their children, for dog meat, or to the far larger number who heard and repeated such stories. What is clear is that several battalions worth of battle-hardened warriors were feeling wronged, and those battalions—despite the promise exacted by Valens, *armed* battalions—were now on the wrong side of the Danube. Diocletian's strategy of defense in depth was about to be tested.

The first battle of the Goth rebellion was fought when the Roman commander at Marcianopolis tried and failed to kidnap and kill two Gothic leaders. The second clash occurred when another group of

Goths, evicted from their winter quarters at Adrianople, 135 miles from the imperial capital, fought an inconclusive battle at Ad Salices, on the western shore of the Black Sea.

By the end of 377, the local troops, the *limitanei*, had fought two battles that, whatever their tactical result, had kept the war from spreading out of the Balkans, and the emperors, Valens in the east, and his nephew Gratian, who had succeeded his father, Valentinian, in the west, had each assembled a field army—two *comitanenses*—to snuff out the problem for good. Despite the logistical nightmare of moving one army from Valens's temporary capital in Antioch, and the other from Italy, the Goths were pinned down where Rome's legions would surely destroy them.

In May of 378, Valens finally reached Constantinople and immediately proceeded west to Adrianople, where he discovered that the anvil against which his army was to hammer the Goths was nowhere to be seen. Gratian had been delayed by yet another of the seemingly endless incursions of barbarian tribes who had crossed into Roman territory that spring. After their defeat at the hands of the western Roman army, the tribes retreated back across the Rhine, tempting Gratian to pursue them in search of a battle that would end his barbarian problems for good and all. In the end, Gratian's choice proved decisive not for the barbarians, but for his fellow emperor, Valens, who would now face the Goths by himself.

Back in Adrianople, as reported by Ammianus, Valens was drowning in choices, each of his advisers recommending different tactics, ranging from waiting for Gratian to show up, to continued skirmishing, to forcing a decisive battle with only a single Roman field army against the Gothic host. Gratian even sent a message asking his uncle to delay the battle until his arrival. For whatever reason—many believe jealousy of his nephew played a decisive part—Valens chose battle.

On August 9, 378, the eastern Roman army, numbering probably 20,000 to 30,000,[12] moved out of the city of Adrianople heading for history. What they found, after hours of marching in stifling heat, was an infantryman's nightmare: at least five thousand archers in close formation, inside a circular palisade. It is not known whether Valens realized that the Goths, like many nomadic peoples, traveled with wagons containing family and possessions, the tongues of which could be yoked together, making for a fortification in record time. Or that Fritigern, the Goth war chief, had sent his cavalry on a foraging expedition earlier in the day, and that the cavalry was therefore not inside the *laager*, but outside it. Whatever the reasons, the right wing of Valens's army, apparently without orders, unsupported by missile artillery (and exhausted from their earlier march in the August heat, amplified by the Goths' decision to set the dry brush of the battlefield aflame[13]) attacked. Unable to maintain the close order necessary for a massed assault, they were thrown back. Only then did they learn that the Gothic cavalry, by one of those fortuitous events that feature prominently in all of military history, was behind them . . . and only when the Gothic cavalry commander ordered the attack, "like a bolt from on high" in the words of the soldier-historian Ammianus, routing "with great slaughter all that they could come to grips with."[14] The charge separated the left wing of Valens's army from its center, and crushed it against the stakes that the Goths had hammered into the ground around their *laager*.

The Roman defeat was total, the worst since Hannibal had surrounded and destroyed eight Roman legions—nearly fifty thousand men—at Cannae in 216 B.C.E. Two-thirds of Valens's army was killed or captured, including the emperor himself, killed by a Goth bowman just after sunset. In Ammianus's telling, the wounded emperor was taken from the field, and placed in a nearby farmhouse just before it was burned down by its Gothic besiegers, thus "unwittingly lighting a funeral pyre for the last Arian emperor of Rome."[15]

After Adrianople communication and travel through the Balkans (which sat astride the only land separating the eastern empire from the west, where Gratian was now in theory in command of all the imperial field armies) was impossible. Not until four years after the battle could the contemporary chronicler Themistius write, "the entire empire shares the same breath and the same feeling like a single organism and is no longer split in two . . ." He would later be among the first to explicitly weigh Adrianople against Cannae on the scales of military disaster, writing that the Goths were "worse for the Romans than Hannibal."[16] Only a month after Adrianople, Bishop Ambrose of Milan gave the eulogy at his own brother's funeral, saying that his brother was taken "lest he should witness the destruction of the entire globe, the end of the world . . ."[17]

The immediate consequences of the debacle were serious indeed; in a single day, Adrianople had separated one half of Constantine's empire from the other, though, to be sure, the separation was incomplete so long as the Mediterranean remained a Roman lake. The real casualty of Adrianople was therefore not Roman commerce, but Roman strategy, since it was far easier to carry grain by ship than troops; moving even a modest-sized army across the Mediterranean could require a thousand ships—two sailors for every soldier. Without the Balkan mountain passes, field armies could not reinforce border troops at anything like an affordable cost.

Over time, however, the fundamental character of the Visigothic military revealed itself to be a tactical asset but strategic weakness: The ferocious army of archers and horsemen that defeated Valens was not only incapable of capturing a fortified city—contemporary chroniclers describe the Visigoths as completely befuddled when confronted by the gates of Constantinople[18]—but could not even control the territories that it occupied.[19] The Goths had learned their own effectiveness as a military force, but also that they would never treat with Rome as equals without a territory of their own. Within a century, their search for such a land would not merely isolate the western half of the empire, but end it forever.

After continuing to fight little-remembered and inconclusive battles with Roman armies from east and west, in 382 the Visigoths entered into treaties that transformed them from the empire's bitterest foes into valued allies—temporarily, anyway, for the *foedus*, or treaty, contained a novel provision that ended a current conflict by guaranteeing one in the future. The specific clause covered the enlistment of Visigothic soldiers into the imperial army, scarcely a new idea. However, while barbarian soldiers had served in Rome's legions for centuries, they had done so as individuals. When those individuals were organized, as called for in the treaties, into cohesive units led by Gothic officers, they became not only the most versatile troops in the eastern army but also the most autonomous. They were, in fact, a virtual state-within-a-state, and therefore a constant worry to Theodosius, who had succeeded Valens as eastern emperor after Adrianople, and became ruler of the entire empire after 392. The worry was well founded. By 395, when Theodosius the Great, "the most just of men and an able warrior,"[20] was succeeded by his sons, the eighteen-year-old Arcadius and his eleven-year-old brother Honorius, the Visigoths were virtually in command of the imperial capital, policing the streets of the city they had been unable to take by assault.

In the words of the historian J. A. S. Evans, "only their restlessness saved the Empire." Perhaps it was nothing more than restlessness of a people whose national myth was explicitly a tale of transcontinental wandering, a set of journeys that would, in time, redraw Procopius's notional map of the European "continent." Perhaps it was the temptation to rebalance the Mediterranean's geopolitical weight after its shift eastward. But whatever the reason, in 395, the Visigoths, pulled westward by the vacuum left by the east-facing prospect of the empire, and led by their newly elected chief, a former general in Theodosius's army named Alaric, began yet another migration across Greece and the Balkan provinces of Pannonia and Illyricum. Their destination was Italy.

The Gothic migrations have always loomed large in any history that attempts to encompass the end of Rome's Mediterranean empire and the birth of what would become the nations of Europe. One of the reasons for the migration's apparent importance, however, is that the Visigoths' journey is seen, across fifteen centuries, through the wrong end of a conceptual telescope. In truth, the Visigoths were nothing like a modern nation, and not merely because they lacked fixed borders, or that they weren't even called Visigoths until Cassiodorus and Jordanes wrote the *Getica*, and concocted a name to distinguish them from the eastern Goths, or Ostrogoths. When the

Visigoths used any word to describe themselves, what they had in mind was closer to "army" than "people."[21] Throughout the fourth and fifth centuries, the Gothic army-on-the-move contained Suevi, Alans, Huns, Germans, and Franks. Goths fought against Huns and with them. It's all rather as if an expedition intended to revenge Custer's defeat of Little Big Horn was led by a Sioux general commanding an army of Irishmen, West Indians, and Chinese.

Still, the impact of this motley coalition-of-the-willing on the formation of modern Europe is large, but subtle. Gothic armies established themselves—sometimes for years, sometimes for centuries—in territories that would one day include Spain, France, Italy, Serbia, and Croatia, among others. As such, they would form the final obstacle in the path of Justinian's planned reunification of the empire. Their migration, and the conflict that was its inevitable result, are two of the indisputably larger forces separating Rome's past from Europe's future.

At the head of their march into Europe's future was Alaric, a member of the Visigoths' hereditary ruling family, the Balthi, and a Goth through and through, despite his training at the hands of the empire. He and his Visigoths may have been drawn westward by the lure of "a plentiful harvest of fame and riches in a province which had hitherto escaped the ravages of war"[22] but their exodus was a paying proposition almost from the beginning, as one Greek city after another gave up its treasure to the pillaging horde. The spoils of Visigothic depredations were not limited to plunder; apparently eager to hasten his unwelcome guest's journey westward, the eastern emperor Arcadius made Alaric *magister militum,* or Master of Soldiers for the province of eastern Illyricum, and the Visigoth wasted no time taking advantage of his new title to arm his troops using the arsenal and arms factories at Thessalonica and Naissus.

Arcadius's brother Honorius, the western emperor, was still only a boy, but he possessed the services of an extraordinarily gifted and ambitious general—possibly part Goth himself—named Stilicho, who could see clearly the danger of allowing the Visigoths an unimpeded path west. In 396, Stilicho led an expeditionary force into western Greece and Macedonia that was able to slow Alaric down with clever maneuvering, but he was either unable or unwilling to bring him to decisive battle, a pattern that would repeat itself so frequently as to generate a widespread belief that the two generals were more eager for cooperation than combat. By early 402, Alaric was besieging Milan, the imperial capital, when Stilicho propitiously turned up with an army that was able to relieve the siege without fighting a battle. Even so, Honorius was frightened enough—if there is a constant in the history of Rome's emperors, it is that the more talented the father, the more disappointing the sons—that he moved the capital to Ravenna, a city on Italy's Adriatic coast.

For the next six years, the Italian peninsula served as the scene of a stalemate between Stilicho and Alaric that resembled nothing so much as the eternal marching and countermarching of fourteenth-century Italian *condottieri* more interested in gold than battle. In 407, the peripatetic Alaric retreated north to what is today Austria, in return for two tons of gold, a sum that Stilicho persuaded the empire to pay with the unassailable argument that four thousand pounds of gold was the annual income of a single Roman senator[23] and therefore a modest enough sum to ensure the security of six hundred of his colleagues. A year later, however, Stilicho's gifts of persuasion were unable to save him from execution at the hands of Honorius, who beheaded his general in 408.* With no Stilicho to bar him, Alaric headed south into Italy, arriving at the "third milestone from the royal city"[24] in 409, where the Roman emperor and Visigothic king met for the first time. According to legend, at that meeting, Alaric challenged Honorius to either live peaceably, or fight a winner-take-all battle for Italy . . . a challenge that Honorius declined, instead suggesting that a more congenial home for the Visigoths might be "the provinces farthest away"[25]—namely, Gaul and Spain. Alaric agreed to withdraw to the north, subject to a payment of five thousand pounds of gold, thirty

---

* That same year, Britain revolted for the last time. It would never again be part of the empire.

thousand pounds of silver, and other trinkets, including a ton and a half of pepper. But either because he was fundamentally untrustworthy, or in response to a treacherous attack by Roman troops, in 410, Alaric and his Visigoths returned. This time, his target was neither Milan nor Ravenna, but Rome.

Even though Rome by now no longer controlled the sinews of the empire, it retained a powerful grip on its heart. Fifteen hundred years later, the sack of Rome remains Alaric's best-remembered achievement, with details accumulating for centuries after the event. Despite all their marching and countermarching from Asia Minor to Italy, the Visigoths still knew more about extorting treasure from walled cities than taking them, and the walls that Aurelian had constructed around Rome in the third century proved a formidable barrier to conquest. Nonetheless, by late summer of 410, the Visigothic army was able to besiege the city. Sixth- and seventh-century historians described Alaric capturing the city using three hundred boys whom he had presented to the Senate as domestic slaves, and who, upon a prearranged signal, killed the guards of Rome's Salarian Gate and admitted the Visigoths. Another, somewhat less cinematic version tells of a Roman matron named Proba, who opened the gates out of pity for the starving citizenry. Either way, on August 24, 410, the Goths entered and plundered Rome, the first time that the city had fallen to an invader in eight hundred years.

While the strategic effects of the sack were inconsequential, and even the direct effects small, the psychological impact of Alaric's victory was monumental, far greater in its way than Adrianople. Less than a hundred years after the followers of Jesus had assumed leadership of the Roman Empire, the fall of the empire's home seemed to strike at the very heart of Christianity. At the other end of the Mediterranean from the birthplace of Jesus, in what is today Algeria, Bishop Augustine of Hippo wrote a tract entitled *City of God* as his answer to those who argued that the fall of the earthbound city was cause for doubting the Christian revelation.

Alaric did not long survive his greatest moment, dying the same year as the sack of Rome. His successor, Ataulphus, led the Visigoths through Southern Gaul—where, in 414, he married Honorius's sister, Galla Placidia—on the next leg of their seemingly eternal migration and across the Pyrenees into Spain. Here their travels ended, as the Straits of Gibraltar presented an obstacle more formidable than the combined armies of the eastern and western emperors; despite decades of travel across Europe, the Visigoths had failed to develop even a rudimentary mastery of the sea. Trapped and starving on the Iberian peninsula, the Visigoths surrendered to Roman authorities, who, somewhat shortsightedly, consented not only to feed them, but to fulfill Alaric's dream of a Gothic *patria* on Roman territory. Honorius agreed to provide the Visigoths with a territorial grant comprising the Garonne valley from Bordeaux to the new "kingdom's" capital in Toulouse, and a strip of the Atlantic coastline from the Loire to the Pyrenees (significantly, no part of the province had direct access to the Mediterranean[26]). In return, Rome required of its new *federates* Visigothic help in cleansing Spain of "other" barbarians.

The shortsightedness was perhaps understandable, given the historical value of the land Rome called Hispania. By the time the Visigoths were hired to police it, Spain had been a province of Rome for nearly seven hundred years,* since the Second Punic War. Conquered by some of the most successful generals in history, including Scipio Africanus, Julius Caesar, and Augustus, it had been the birthplace of dozens of the empire's greatest figures—Trajan and Hadrian, the first two emperors from Rome's provinces, were both born in Spain, as were Theodosius and the writer Seneca. Its residents were the first of the empire's provincials to automatically receive Roman citizenship, and its rich lands were part of more than one great senatorial fortune.

Spain's wealth attracted more than greedy senators. While Alaric was crossing swords with Stilicho, it caught the attention of a barbarian people, originally from the area around the Sea of Azov, known to history as the Vandals. The Vandals had been living in Gaul, more or

---

* Actually, three provinces: Tarraconesis, Lusitania (most of modern Portugal), and Baetica in the south.

less peaceably, since approximately 400; but when they, along with other tribes, including Alans, crossed the Pyrenees into Spain in 409, they were treading dangerously close to the empire's heartland. Using one barbarian army to defeat another might have seemed the best of a bad set of options to Honorius, still ruling the western empire from his palace in Ravenna; but in the event, the Visigothic strategy was successful, as far as it went. When the Visigoths were ordered back to Gaul on the death of Honorius in 425, Spain was largely pacified, with the Alans either decimated or absorbed into the Visigothic army, the Vandals trapped in the south, and the Suevi, another barbarian tribe, largely pinned on the western side of the Tagus River, in what is today northwest Spain and Portugal.

In return, however, Rome now had to contend with a Visigothic kingdom that was effectively ruling southern France and much of Spain from Toulouse. Worse still, Spain's remaining Vandals, eighty thousand strong,[27] crossed into Africa in 429 and ten years later, under their king, Gaiseric, they took Carthage itself. Before signing a peace treaty with Rome in 442, the Vandals, who, unlike the Visigoths, were skilled sailors, regularly raided throughout the Mediterranean, pillaging even Sicily and Apulia. The western half of Constantine's empire was rapidly eroding to little more than the portions of the Italian peninsula ruled from Ravenna. The erosion would not, however, end there.

―――――――――

By the middle of the fourth century, the Ostrogothic kingdom covered an enormous expanse, from the Dniester (the border between Ostrogoths and Visigoths) to the Don in the Ukraine, and from the north shore of the Black Sea to what is now Belorussia. Their easterly position granted them an early encounter with the Huns, who conquered them in the early 370s.

Given their brief appearance on history's stage, and the contempt with which contemporaries described them—"Like unreasoning beasts, they are utterly ignorant of the difference between right and wrong; they are deceitful and ambiguous in speech . . . [and] are so fickle and prone to anger that they often quarrel with their allies without provocation. . . ."[28]—the Huns had a remarkable and long-lasting effect on both the Roman Empire and the European nations that would appear after its fall. Their conquest of the eastern Goths was only a prelude to their first appearance in force south of the Danube in the winter of 394–95, a winter in which the frozen river served as a bridge to the Hun cavalry. By 425, Constantinople was paying Huns the rather paltry sum of 350 pounds of gold a year as protection money (as comparison, recall that Alaric had been able to extort 4,000 pounds of gold to leave Illyricum in 407). By 434, Rua, the king of the Huns, was treating with Constantinople as an equal, demanding diplomatic niceties like an extradition treaty that would permit the return of Huns who had rebelled against him. Rua was formidable enough that the entire eastern empire was likely relieved when, in 434, he died. If so, its relief was to be shortlived, as the Huns chose as their new king a warrior named Attila.

Priscus of Panium, a member of the Roman negotiating team that attempted to forge a peace treaty with Attila in 449, described him as "short and stocky, with deepset eyes in an oversize head, and wearing a thin beard."[29] In the words of the Roman historian Renatus Frigeridus (probably depending on secondhand reports), the Hun chieftain was "quick of wit and agile of limb, a very practiced horseman and a skilful archer . . . a born warrior, he was renowned for the arts of peace, without avarice and little swayed by desire, endowed with the gifts of the mind, not swerving from his purpose for any kind of evil . . ."[30] His terrifying warriors regarded him as the most formidable among them, a man as much feared in single combat as in planning a cavalry charge; his opponents used his name to frighten children—a purpose it still serves today. But despite his well-earned reputation as the Scourge of God, Attila was a patient negotiator, "temperate in all things,"[31] who achieved as much by diplomacy as by his military prowess. Thus, while the new Hun king was regularly crossing the Danube on smash-and-grab raids into Constantinople's sphere of influence throughout the 430s, the goal of the raids was less the treasure pillaged than the protection money extorted; Constantinople raised the annual subsidy first to 700 pounds of gold, then to 2,100.

In 447, Attila again crossed the Danube, and again defeated each of
the Roman armies sent against him. But this time, out of boredom or
simply prudence—possibly, Attila had decided that the east had paid
as much as it was going to—the Huns sent some exploring parties
westward.[32] In 450, one of them returned, with a letter.

The sender was Honoria, sister of the western emperor, Valentinian III, who had succeeded his uncle, Honorius, in 425. Her letter
to Attila, which informed the Huns' king of her distaste for the husband selected for her by her brother, included her ring, which gave Attila the pretext he needed to march into Gaul, claiming that he and
Honoria were betrothed. For the next three years, the entire western
empire was besieged by Hun armies, comprising dozens of conquered
peoples, such as Alans, Ostrogoths, Suevi, even Vandals . . . as did the
armies that Rome fielded against them, most memorably in 451 when
a combined Roman and Visigoth army defeated the Huns' army
somewhere near what is today Chalons in central France. Or, rather,
they defeated Attila's army, for unlike Rome, the Huns' "empire" was
almost entirely the creation of a single remarkable man.

Remarkable, but not immortal. In 453, on the night of his wedding
to the latest in a string of wives, Attila suffered a nasal hemorrhage
and drowned in his own blood, a metaphorically rich end for the
Huns' ruler. His funeral song, chanted by the army's finest horsemen
as they circled his corpse, was recorded by Priscus:

> Chief king of the Huns, Attila, son of Mundzuc, lord of the bravest
> peoples, who possessed alone the sovereignty of Scythia and Germany
> with power unheard of before him and who terrorized both empires
> of the city of Rome by capturing their cities and, placated by their
> prayers, accepted a yearly tribute lest he plunder the rest. When he had
> achieved all these things through his good fortune, he died not by an
> enemy's wound or through treachery of his followers, but painlessly
> while his people was safe and happy amidst his pleasures.[33]

The collapse of the Huns as a political force upon the death of Attila
remains somewhat puzzling, since they had terrorized great swaths of

imperial territory for years before his accession, and remained fear-some warriors for centuries after his death. As with all the other migratory armies that flowed over fourth- and fifth-century Europe, the Huns' ability to confront the empire was limited by their lack of a permanent home with defensible borders. Absent the defensive and economic advantages of such a sovereign territory, the Huns could raid—and, under Attila, terrify—Rome, but lacked any larger war aims, and when Attila's armies were divided among his sons, the po-litical power of the Huns vanished.

The same could not be said for the Huns' tributary nations, so while Rome's emperors cheered the immediate consequences of At-tila's passing, the long-term sequelae were considerably less happy. At-tila's victories, as many have observed, were far less dangerous to the empire than his death[34] since the latter permitted the Vandals, Alans, Suevi, and Burgundians to act in their own interests, and they wasted little time in doing so.

The result was the final and complete unraveling of the western empire. Two years after the Hun king died, Rome was sacked again, this time by Gaiseric's seagoing Vandals, whose new North African home enabled them to raid Italy without fear of reprisals. In 455, no fewer than three emperors died—two by assassination, one in a riot. The last, Avitus, was deposed by a Suevian general named Ricimer who had a gift for emperor-making that he was to exhibit for the re-mainder of his life.[35]

Perversely enough, Ricimer might have served Rome better had he respected her less. Like Stilicho before him, the Suevian has mystified generations of historians by not striking for the throne himself, a testi-mony to the powerful hold that Rome—or, more precisely, the Ro-man *idea*—still retained. Unable to persuade himself that he could rule in his own name—Ricimer was both part-Visigoth and all Arian; the first non-Roman to accept the title of emperor in the west was Charlemagne, in 800[36]—he instead placed a series of others on the western throne, each less vigorous than his predecessor, and most of whom were eventually discarded by Ricimer by the time-honored means of imperial power transfer: death by poisoning.

The impotence of Ricimer's comic-opera emperors was mirrored in their domain; during the five decades since the Goths settled in Toulouse, the western empire had lost control, successively, of Africa, Spain, most of Gaul, and, finally, of the ability to command its own army.[37] The barbarian _foederates_ were the only formidable troops in the whole of Italy, and, in 476, they demanded their own territory, as had been given to the Visigoths in Gaul. When the latest entry in the register of western coemperors—Orestes and his son Romulus Augustulus—refused, the federates rebelled, naming a new leader, Odoacer, who, in the year 476, killed Orestes, and deposed Romulus. In an act less like conquest than mercy killing, the reign of the last Roman emperor in the west had ended.

Perhaps the most significant consequence of the death of Attila—considerably more important than the overthrow of Romulus Augustulus—was its effect on the Ostrogoths. Until Attila's passing, the Ostrogoths, ruled by the Amal family as Hun vassals, had remained in their Black Sea home, while their Tervingi–Visigoth cousins were migrating west. Freed from their Hun masters, the senior members of the Amali family began jockeying with one another for ascendancy over the Ostrogoth protonation.

They were still jockeying eighteen years later, when an imperial-trained Amali soldier named Theodoric Strabo—Theodoric "the Squinter"—emerged as the most powerful "generalissimo" in the east. Self-proclaimed as the king of Thrace in 473 and leading a formidable army loyal only to him, Strabo began to look to the eastern emperor, Leo, a lot like Odoacer: namely, a barbarian powerful enough to push an emperor from his throne. In order to pacify him, Leo negotiated a recognition of Strabo that granted him nominal authority over all the Goths remaining in the east.

When Leo died a year later, however, he was succeeded by a more belligerent emperor, Zeno, who abrogated Leo's treaty of recognition almost immediately, believing that the most effective way to curb the Ostrogoth threat was not to bribe them, but to divide them. As it happened, despite the imperial warrant, Strabo did not command the loyalty of much outside his own Thracian army, and the other Ostrogoths, and *their* Amali leaders, were disposed to help the empire, if it meant helping themselves. Zeno needed an Ostrogoth king as a counterweight to Theodoric Strabo, and found him in his own army.

The "other" Theodoric was born in the early 450s into the royal Amal family, and spent his boyhood in Constantinople as a hostage against the good behavior of his father, an Ostrogoth king named Theodemir. In 474, he succeeded his father to the title, but the kingship meant little in practical terms until 476, when Zeno appointed him to the Patrician Order (an honor that still had both cultural and legal advantages within the empire). More important, he made him the commander-in-chief of an Ostrogothic army, which he subsidized in order to frustrate the ambitions of Strabo. The Ostrogoths now had, de facto, two kings, both named Theodoric, both members of the same family, fighting each other.

For the next five years, Zeno played divide-and-rule games with the two Amali kinglets, bribing one, lying to another, throwing imperial troops into the occasional battle, shifting support whenever one Theodoric looked to be gaining a decisive advantage over the other. By the time Theodoric Strabo died, in 481, the younger Theodoric had "survived the greatest crisis of his career,"[38] learning, along the way, the tactics that would earn him the appellation the Great.

Part of that greatness, to be sure, is the beatified version of Theodoric's life and ancestry bequeathed to subsequent generations by Jordanes's *Getica*, based as it was on the work of the Ostrogothic king's house panegyrist, Cassiodorus. Even so, the facts about Theodoric's life after 481, and the regard in which he was held by other contemporaneous historians, are impressive enough. Certainly, Zeno was impressed by the now-unified Ostrogoth army, not to say frightened. When the emperor started to fear his employee as much as the people against whom he was employed, he looked for a target to distract the Ostrogothic army. He found it in the west.

The explicit offer to the Gothic warrior was that of *magister militum*, Master of Soldiers in Italy, to which Theodoric repaired, in 489, leading a motley but formidable army of Gepids and Burgundians as well as his Ostrogoths. After four years of bloody though inconclusive battle, he had Odoacer besieged in Ravenna, where he had him killed by subterfuge in 493. That same year, Theodoric was acclaimed king.

By 510, the Ostrogoth king ruled all of Italy and portions of the Balkans, putatively as representative of the eastern emperor, really as an autonomous monarch. Theodoric's kingdom was, in retrospect, an unstable mixture of Gothic Arianism and Italian orthodoxy, held together by Theodoric's enormous personal magnetism. But its coherence is also a vivid illustration of the still-powerful centripetal attraction that Roman civilization still exhibited against the forces pulling that civilization apart. Like Ricimer, Theodoric maintained a healthy respect, not to say awe, for all things Roman, famously writing, "We delight to live after the law of the Romans . . . for what profit is there in having removed the turmoil of the Barbarians, unless we live according to law?"[39] The watchword of the Ostrogoth's reign was *civilitas,* understood by Theodoric (wrongly, though widely, believed to be illiterate) as the defining aspect of civilized life; one of the opinions in his *Edict,* the recodification of 154 Roman laws that applied equally to Goths and Italians, read: "We do not love anything uncivilized."

Theodoric's need for Roman legitimacy was also political. As early as 489, when he first entered Italy, he had asked Zeno, his erstwhile patron, for the right to rule there as a king. But by 490, Zeno was dead, and his successor, Anastasius, selected by Zeno's widow, Ariadne, as her new husband, turned Theodoric down. Perhaps the new eastern emperor felt the shade of Constantine looking over his shoulder, disgusted by what had become of his empire in the two hundred years since his death. Italy, Spain, and North Africa had been the core of the Roman state ever since it earned the name empire. By the time of Zeno's death, Italy was ruled by Ostrogoths, Spain by Visigoths, and North Africa by Vandals; and Constantinople was to be ruled by Anastasius. He was an unlikely choice. Zeno had been a formidable warrior, and relatively unlettered; Anastasius was a lay preacher and theologian, sixty years old, and lacking in any military credentials. Practically the only thing the two men shared, besides Ariadne, was a particular brand of Christian philosophy called Monophysitism.*

In Constantine's time, the great theological dispute pitting Arian Christians against orthodox turned on the relationship between God-the-father and God-the-son. A hundred years after the Council of Nicaea signed off on the doctrine of consubstantiality, some were still teaching that though Jesus died as God (or, at least as the divine Logos), he was born and suffered as a man. Logically, then, the honorific "Mother of God," which had long been granted to Mary, was false. This was the conclusion of Nestorius, Bishop of Constantinople from 428–431, and of the Nestorians, his followers.

And this was a problem. Even by the third century, worship of Mary was a well-established feature of the Christian world, and the uproar caused by Nestorianism caused its founder to be deposed and exiled at the Council of Ephesus in 431, where Mary was named *theotokos;* i.e., mother of God. Ever since, the divinity of Jesus has been the very heart of orthodoxy. For many, however, particularly in the eastern provinces of the empire, Ephesus did not go nearly far enough. An Alexandrian priest named Eutyches staked out the extreme corollary of the orthodox position: that the nature of Christ contained *no* difference between human and divine, only a single-natured—Monophysite—Jesus. This doctrine was ratified at the second Council of Ephesus, where a packed court of eastern priests, largely from Alexandria and Antioch, strongly supported the Monophysite position, which was just as strongly opposed in the west by both Arians (who contended that Jesus was at least *part* human) and traditionalists (who believed in Jesus as *both* god and man).

To a post-Enlightenment reader, such debate seems trivial indeed,

---

* Many modern scholars are not fully comfortable with the term "Monophysitism," preferring "Miaphysitisim" or even "anti-Chalcedonianism." Partly, this is simple courtesy toward some modern religious groups, including Egyptian's Coptic Christians and some Syrian and Armenian churches, who profess the doctrine but who do not call themselves Monophysite. With apologies, however, this book will continue to use the more familiar term.

just the sort of pointlessly ornate reasoning from which the word "byzantine" gets its dictionary definition. To understand the passions with which matters of theology were viewed in late antiquity—to be sure, not by everyone; Constantine himself was dismissive of such thinking—it is necessary to engage those passions on their own ground. The ground is unusual, if not unique, in the long history of mankind and its religions; neither Judaism nor Islam (to say nothing of other, non-Western religions) has a "theology" in the way that Christianity does; that is, a body of knowledge concerning the nature of God.

Christianity's unique need for explaining the nature of God is a direct consequence of the paradox intrinsic to what had been, since the demise of the heresy known as Gnosticism,* the central element of Jesus's life: the ending of it on the cross. Reconciling the suffering death of an omnipotent God is the distinctive challenge of Christianity, expressed in a wide array of profound questions, some still asked today: If Christ was not merely the Messiah, but in some sense God, how is it that he had no recorded existence before his earthly one? Why was there no mention of Jesus in Eden, or with Noah at the Flood? Where was Jesus when Abraham and Moses spoke directly with God? Such concerns mattered little to the power of Christ to attract believers, frequently poor people who cared less about intellectual coherence than emotional resonance. The men who ministered to those believers, however, remained obliged to generate and respond to intellectually sophisticated attempts to square the circle, to explain how, even though Jesus is the son of God, he was not subordinate to, nor even equal to, but in some ways the *same being* as his Father.

Such questions about doctrine, inherent as they are to the unique character of Christianity's founding tradition, were probably unavoidable, and the leaders of the early Church, most prominently Peter and

---

* Gnosticism is a powerfully complicated worldview, covering much more than the decision to give preeminence to Jesus's life and acts over his death. In some form, it even precedes Christianity itself, in the form of arguments that the presence of evil in the world is evidence for creation by an entity other than God.

Paul, spent their lives debating Gnostics over them. Monotheistic religions are famously intolerant of apostasy, even when they disagree about what constitutes it. The disputes, largely a matter of whether to emphasize Jesus's wisdom (in Greek, *gnosis*) or his suffering, did not, however, become politically important until Christianity became the de facto state religion of the world's greatest empire, with all that that entailed. Once Christianity was identified with the Roman state, doctrinal struggles within the former were inextricably bound up in the governance of the latter, and any real dispute could only be settled by both temporal and spiritual authorities. The settlement, when it came, was a total defeat for the Gnostic tradition, and what would come to be called orthodoxy—in Greek, "straight thinking"—became dependent on the savior as a sacrifice, rather than a teacher, and *that* meant that His death was at least as important as His life.

In Rome, Pope Leo, in his *Tome* of 449, articulated the idea that Jesus personified two natures: "he that is true God is true man . . . the Word performing what belongs to the Word, and the flesh carrying out what belongs to the flesh. The one sparkles with miracles, the other succumbs to injuries."[40] The obvious implication was lost on no one: Just as Christ had two natures, one human and one divine, with both necessary for a *single* godhead, so does the world have both a pope and an emperor . . . and that which affronts one attacks the other. The year after Leo's *Tome*, the eastern emperor Marcian persuaded Leo to agree to call yet another council at which the pope's own doctrine, Dyophysitism, could be established. Thus, on October 8, 451, the Council of Chalcedon decreed that orthodoxy required believers to acknowledge that Christ was of "two natures, without being mixed, transmuted, divided, or separated."

As is often the case with doctrinal disputes, however, what looked to one side like a compromise appeared to the other as a provocation. Fourteen hundred years later, John Henry Cardinal Newman would write, "a doctrine . . . which the whole East refused as a symbol, not once, but twice, patriarch by patriarch, metropolitan by metropolitan . . . was forced upon the Council . . . for its acceptance as a definition of faith under the sanction of an anathema, forced on the Council by the resolution of the Pope of the day, acting through his Legates and supported by the civil power."[41] He did not exaggerate: In 452, when Monophysites were forbidden to have priests or to own property, the monks of Palestine raided Jerusalem and evicted the bishop; Egypt's Monophysite priests rioted, and when the emperor's soldiers attempted to stop the riot, burned them alive in the Temple of Serapis.

By the time Anastasius became emperor in 491, both the Monophysite and Chalcedonian positions were being expressed somewhat less violently than forty years earlier, but the east-west lines along which the dispute had formed had hardened. Anastasius, though Monophysite himself, was relatively tolerant of religious disagreement, so much so that his orthodox opponents in Rome mistook his tolerance for weakness. In this, they were quickly disabused; in 517, when a delegation from Pope Hormisdas ordered Anastasius to remove the name of the Monophysite-leaning patriarch of Constantinople, Acacius, from public displays, the emperor famously replied, "You may insult me; but you may not command me."[42]

The studied hauteur of the emperor's reply is that of a confident ruler without any unmet ambitions. He had reason for his confidence. Though Anastasius was possessed of some quirks—the puritanical ruler not only tried to ban animal fights from the empire but also pantomime[43] (the latter being not the white-faced twentieth-century street theatricals, but explicit pornography)—his twenty-seven years of rule left the empire extraordinarily prosperous, mostly by attending to the unglamorous side of administration, efficiently calculating, for example, how much tax to collect in kind, and how much in gold. In what is almost certainly the first documented exercise of what would come to be called trickle-down economics, Anastasius abolished a wide range of taxes that fell heavily on the empire's most productive classes, its craftsmen and merchants. The emperor had argued, it turns out correctly, that a prosperous merchant would pay even more in fees than the treasury lost in taxes. Thus, despite three major wars, and several revolts by subjects opposing the emperor's Monophysitism, the

treasury at Anastasius's death was richer by 320,000 pounds of gold*
than it had been at his accession.

Of the few things left undone by the eighty-seven-year-old Anastasius when he died on the night of July 9, 518, one loomed large: He
had failed to name a successor. A series of rebellions fueled by resentment of the emperor's Monophysitism was fresh in the minds of Constantinople's elite when the Senate convened, a day later, less than a
mile from where a restive crowd filled the city's great Hippodrome. At
the end of a day's worth of intrigues—supposedly one of the candi-
dates for Anastasius's throne gave a guard commander money with
which to bribe his troops; he did, but on his own behalf—a compromise candidate was selected.

Anastasius's successor, the inheritor of his 120 tons of gold, was a
throwback to his predecessor. Zeno had been widely known as an unsophisticated warrior; the new emperor was a soldier, the commander
of the *excubitores* regiment, whose career, admirable enough on its
own merits, was destined to remain little more than a footnote to that
of *his* successor: Justinian, the last Roman emperor to earn the title
"the Great."

---

* Twenty-three million *solidi*. Among Constantine's achievements was the establishment of a
new standard for imperial coinage, replacing the *aureus* with the somewhat lighter *solidus*,
seventy-two of which made a twelve-ounce "pound" of gold.

CHAPTER THREE

"Our Most Pious Consort"

518–530

THE CIRCUMSTANCES OF Justin's accession to the imperial throne were so improbable that they practically demand a hidden hand stage-managing the process: spreading rumors, bribing senators, instigating "spontaneous" demonstrations of support among the crowds in Constantinople's streets. Since the beneficiaries of the compromise decision were two Illyrian peasants, and no one, then or now, ever deemed Justin a skilled politician, most historians have assumed that the hidden hand belonged to Justinian. If he indeed snatched an empire for his uncle, he also helped him to rule it.

For nine years, Justinian served as de facto vice-emperor—and, as Justin exhibited more and more signs of senility, sole ruler. During that time, the regime's most urgent tasks were religious; a large, probably decisive, component of Justin's appeal to the senatorial aristocracy and peasantry alike was his loyalty to the creed that had been accepted as orthodoxy at Chalcedon. Reconciling with Pope Hormisdas required that the new emperor repudiate his Monophysite predecessors Zeno and Anastasius, and rid Constantinople of the clerics that they had summoned to the capital. By 521, when Justinian sponsored a series of games in celebration of his first Consulate (a still prestigious honor hearkening back to the Roman Republic, and an opportunity to secure the political favors that are the stock in trade of any politician) the Monophysite patriarch Timothy had been sent back to his home in Alexandria, and Constantinople was as orthodox as Rome itself.

Rather more so, in fact, since Italy was by then ruled by an Arian heretic, the Ostrogoth king Theodoric the Great. Nonetheless, since the Ostrogoth, in theory, ruled at the pleasure of the emperor, Justin agreed to a symbolic "adoption" of Theodoric's son-in-law Eutharic, and granted the members of the Ostrogothic court a number of prestigious imperial titles. There is little doubt that Theodoric, who famously wrote "an able Goth wants to be like a Roman; only a poor Roman would want to be like a Goth,"[1] valued recognition by Constantinople's emperor, even though his own authority extended over a domain nearly as large, and despite a predictable wariness about imperial policies. Ever since 507, when his son-in-law, the Visigothic king Alaric II, was killed in battle against the Franks, Theodoric had ruled Spain as regent for his grandson, as well as the entire Italian peninsula and portions of the Dalmatian coast. When Constantinople's emperor—or, more precisely, Justinian—looked to the west, to the central territories of the realms of Augustus and Constantine, he saw it covered by a large Ostrogothic shadow. The shadow was essentially that of Theodoric himself, which made any reassertion of Constantinople's authority hostage to his health, and that of his successor. No doubt Theodoric realized this, for when his son-in-law Eutharic died in 522, and his friend and ally Pope Hormisdas* the following year, the Ostrogoth king's caution grew into a paranoid episode that would lead directly to the creation of a work that, more than any other, marks the intellectual doorway between the classical world and the medieval world that would replace it.

Anicius Manlius Severinus Boethius was born about the same time as Petrus Sabbatius Justinianus, but in circumstances that could not have been more different. A wealthy member of one of Rome's oldest families, the Anicii, Boethius was multilingual, a mathematician, philosopher, and rhetorician, and a representative of the very highest of Roman culture during late antiquity. With the ascendance of an Ostrogothic king with a passionate affection for all things Roman, Boethius's future as a courtier was made; indeed, both his sons were to be named by Theodoric as consuls for the year 519. In 522, however, a

---

* The friendship between the Arian king and the orthodox pope—the same pope who scolded Anastasius for his Monophysite sympathies—remains an eyebrow raiser.

senator named Albinus was accused by Theodoric of conspiring with Justin; and when Boethius spoke in the defense of his senatorial colleague, saying in effect that to suspect Albinus was to suspect the entire senate, his king took him at his word and Boethius was jailed, tortured, and executed in October of 524. Before he died, however, he authored one of the great works of antiquity, the Neoplatonist classic *The Consolations of Philosophy*.

In the *Consolations*, Dame Fortuna, the personification of the opaque and mysterious power of fate, was regularly invoked as a—perhaps *the*—force driving human history. That a devout Christian like Boethius would, in extremis, invoke Plotinus rather than Jesus, is ample proof of the ways in which Hellenism retained its hold on the Mediterranean world. Boethius's great work—the "golden volume" as Gibbon called it, no doubt favoring any text that made as little of Christianity as did Boethius—is one of the truest bridges to the Middle Ages that can be found in antiquity. It was translated into virtually every European language, most famously into English by Geoffrey Chaucer, who was likely attracted less by the work's philosophical rigor than the poems that break in regularly "like organ chants in a religious service."[2] Probably the most vivid and recalled image from the *Consolations* is that of Fortuna, whose famous wheel randomly allocates the material wealth of the world without regard to its recipients' virtue. Less well remembered, perhaps, is the eponymous "consolation" offered by Philosophia: "Nature has not quite abandoned you. In your true belief about the world's government—that it is subject to divine reason and not the haphazards of chance—lies our greatest hope of rekindling your health . . . and you will be able to see the resplendent light of truth."[3]

Boethius did not write the *Consolations* for any audience save himself, and certainly not for the attention of the emperor in Constantinople; his attention, however, is indirectly responsible for its existence. That Theodoric would sanction the torture and execution of his own closest adviser is, in its way, a measure of the danger that he saw looming over the eastern horizon, a force fundamentally hostile to his dream of an Ostrogothic *patria* in Italy. The force, which would turn the entire Mediterranean upside down in the years between the death of Theodoric in 526 and the arrival of the demon in 542, was neither geography nor history, but the ambition and personality of the emperor-in-waiting, Justinian.

————

Justinian's temperament is as complicated as his legacy, and—to some—as ambiguous. About his productivity and industriousness there is little disagreement. Insomnia was only the most obvious example of his obsession with work and extraordinary capacity for taking pains: One of the laws with which he concerned himself directly covers a regulation of the price that Constantinople's small farmers would be permitted to charge for their fresh vegetables. Like Richard Nixon, another insomniac introvert who was insecure his entire life around men of better background or easier grace, Justinian possessed a brilliant intellect capable of grand dreams and petty suspicions within the same day—even the same hour—and showed the "ill-disciplined energy of all insomniacs as he paced the palace corridors at night . . . an intellectual who thought big thoughts and had a cosmic sense of his own destiny . . ."[4]

Likewise, everyone who has written about him is in agreement as to his appearance and dress, both of which were as commonplace as his mind was exceptional: middle height, neither fat nor thin, round face, sharp nose, receding hairline.[5] According to Procopius, the great historian of Justinian's reign, "His temperance was ascetic. In Lent he used to fast entirely for two days, and during the rest of the season he abstained from wine and lived on wild herbs dressed with oil and vinegar."[6]

On more substantive issues, however, contemporary descriptions should be approached with caution: One sixth-century chronicler wrote,

> Justinian was insatiable in the acquisition of wealth, and so excessively covetous of the property of others, that he sold for money the whole body of his subjects to those who were entrusted with offices or who were collectors of tributes, and to whatever persons were disposed to

entrap others by groundless charges. He stripped of their entire property innumerable wealthy persons, under colour of the emptiest pre-
texts. If even a prostitute, marking out an individual as a victim, raised
a charge of criminal intercourse against him, all law was at once rendered vain, and by making Justinian her associate in dishonest gain,
she transferred to herself the whole wealth of the accused person.[7]

This account reports what was probably a common understanding
among the wealthier citizens of Constantinople, by default the class
whose opinions are most likely to make it into the permanent historical record. Relying exclusively on them for a clear picture of the emperor is rather like writing a biography of Theodore Roosevelt using
only the diaries of J. P. Morgan.

Luckily, an understanding of Justinian is not dependent either
upon the writings of his contemporaries nor the image appearing on
the coins struck during his reign. Unusually for an emperor, Justinian
was a writer in his own hand, and he did not write to obscure, but to
reveal. Much of what he chose to reveal was a taste for, and mastery of,
Christian doctrine, and the more time a modern reader spends in the
company of the emperor's mania—there is no better word—for theological disputation, the more exotic it seems. The only real modern
analogue is early-twentieth-century Bolshevism, another political culture that demanded mastery of a complicated official ideology. But
Justinian seems to have immersed himself in Christology not to improve his political prospects, but out of a sincere, if occasionally
clumsy, interest—though even the word "interest" betrays a set of
post-Enlightenment blinders. Justinian's religion was not a hobby he
had chosen, but a faith. Justinian, and his contemporaries, subscribed
to a belief system that didn't need to define itself against forces like science, or humanism, or dialectical materialism, and the honest assessment of it is to recognize it as an artifact of an age that is not, finally,
fully comprehensible to a modern.

Not all aspects of Justinian's character are quite so exotic. His writings reveal an enormous ego—he is distinctly wanting in respect for

his predecessors, up to and including Augustus himself—an extraordinary command of detail, and a brutal and decisive hostility to everything that he considered an enemy of the Church, whether from within—Donatists, Nestorians, and Arians—or from without, such as pagans and Jews.* In *Deo Auctore* (by God's Authority), Justinian's imperial constitution, he writes:

> We govern under the authority of God our empire which was delivered to us by His Divine Majesty, we prosecute wars with success, we adorn peace, we hold up the framework of the State, we so lift up our mind in contemplating the aid of the Omnipotent God, that we do not put our faith in our arms nor in our soldiers, nor in our leaders in war nor even in our own skill, but we rest all our hopes in the providence of the Supreme Trinity and in Him alone, whence have proceeded the elements of the whole universe and by whom their disposition throughout the earth's globe was planned.[8]

Justinian was an innovator and reformer who would have been mortally insulted had anyone ever referred to him as such, a man only able to modernize "by convincing himself and others that he was restoring the past . . .";[9] the idea of innovation, to a sixth-century Christian, was virtually heretical.[10]

He was also the least peripatetic of men. He left his capital city seldom, if at all, except to visit nearby vacation palaces. He did not inspect bishoprics, visit officials, or campaign with the army. Rarely in history has a great conqueror showed less interest in visiting his conquests . . . and, indeed, why would he? He had Constantinople.

————

The city that had welcomed him when he arrived with Illyrian dust still caking his sandals never lost its allure during his ascent to its highest place. Constantine's capital might have been built with Justinian in mind. The metropolis was probably the one spot on earth that could satisfy the onetime provincial's two great desires—the exercise of

* Significantly, his antagonism to Monophysitism seems to be irregular at best.

political power, and sophisticated debate about religious dogma. While it was a giant chessboard of broad boulevards and open squares leading to dozens of palaces and churches, the city's center of gravity was a house not of liturgy or law, but of leisure.

In today's Istanbul, what remains of the Hippodrome occupies a long plaza, with the great obelisk at its center thrusting up like a chimney that is all that is left of a fire-consumed house. Surrounding it, more or less along the track once taken by the great charioteers, are grassy rectangles laid into the earth like enormous mosaic tiles. From the base of the obelisk, one can easily see the city's great church, though it would have been invisible to the Hippodrome's occupants. Now gone, its great walls, eighteen meters high, more than one hundred meters wide by five hundred meters long, directed all eyes inward, to the contests staged there a hundred times a week.

During a chariot race, by far the most popular of the competitions staged at the Hippodrome—gladiatorial contests had been banned by Constantine in 326[11]—more than seventy thousand people crowded onto the arena's benches. A large number, assuredly, were there simply for love of the sport, which had been attracting fanatical devotion for six centuries. But many, if not most, were there to exhibit their support for one faction or another, and the politics of the empire would, more than once, be determined by that support.

The racing factions themselves dated back to the first century B.C.E., when four were recorded, identifying themselves with the colors worn by their favored charioteers: red, blue, green, and white. The factions maintained stables; recruited, trained, and paid charioteers; and booked bets on the race's outcome. During the days of the Roman Republic, Pliny describes the suicide of a fan during the funeral for a much-loved Red charioteer, and no chronicle of the early empire, from Suetonius's *Twelve Caesars* to the *Satyricon* of Petronius, fails to mention them.[12] By the time of Justinian, the Reds had combined with the Blues, and Whites with Greens, leaving two factions whose interests had expanded far beyond the results of the Hippodrome's races. The Blues and Greens competed for political advantage not only by supporting their respective horses and charioteers, but rhetorically.

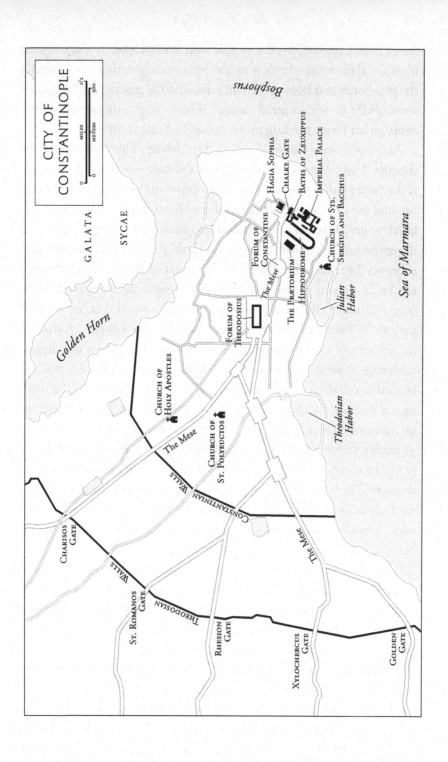

CITY OF CONSTANTINOPLE

MILES   1/2
METERS   500

Bosphorus

GALATA

SYCAE

Golden Horn

Sea of Marmara

Hagia Sophia
Chalke Gate
Baths of Zeukippus
Imperial Palace
Forum of Constantine
The Mese
The Praetorium
Hippodrome
Church of Sts. Sergius and Bacchus
Julian Habor

Forum of Theodosius

Church of Holy Apostles

Church of St. Polyeuctos

The Mese

Theodosian Habor

Constantinian Walls

Charisos Gate

Walls

St. Romanos Gate

Theodosian Walls

Rhesion Gate

Xylochercus Gate

Golden Gate

The Mese

As early as Constantine, the people were encouraged by the emperor to make their wishes known to the provincial governors by cheering the good ones and booing the bad ones, and the practice had migrated successfully to the imperial capital. There, slogan-shouting had become an art form, needing professionals to bring it off.[13]

And not merely slogan-shouting, but debate. The *Chronicle* of the courtier Theophanes faithfully records a debate—perhaps disputation is the better word—between Justinian (through his herald, or *mandatus*) and the chosen representative of the Green faction. The dialogue is startling on a number of grounds. First, the Green "debater" addresses the emperor, the viceroy of Christ on earth, practically as an equal. He addresses Justinian respectfully—as "Justinianus Augustus"—but registers his complaint precisely as if he were doing so before a small claims court, informing the most powerful man in the world that "my oppressor can be found in the shoemaker's quarter." For his part, Justinian, though clearly aware that he holds what might be called a preemptive advantage ("Verily, if you refuse to keep silent, I shall have you beheaded"), still debates both the truth of the Green claims and the theological position that he suggests informs those claims. Justinian tells his interlocutor, "I would have you baptized in the name of one God" only to receive the response, "I *am* baptized in One God," evidently an attempt to contrast his Monophysite sympathies with the emperor's orthodoxy. The Green spokesman accuses the emperor of suppressing the truth, of countenancing murder, and when he has had enough, he ends with "Goodbye Justice! You are no longer in fashion. I shall turn and become Jew; better to be a pagan than a Blue, God knows . . ."[14]

The most telling part of the entire dialogue, however, is that it was entirely conducted in a rigid metrical form with each call-and-response containing the same number of Greek syllables, and placing the stress on the penultimate or antepenultimate syllable for each line. That this was apparently a commonplace occurrence in the Hippodrome demonstrates both the level of training and theological sophistication required for such poetic improvisation.

In Constantinople, the road that led from formal debate to sloganeering to abusing officials to street violence was a short one. By the time of Justin's accession, the Blues and Greens each had shock troops, called Partisans, who resembled nothing so much as modern urban street gangs. They sold contraband, extorted money from merchants, vandalized the city, and raped women. The Partisans wore color-coded uniforms consisting of short blue or green coats, in the style of the Huns, rather than more modest Constantinopolitan tunics.[15] As a young man, Justinian had recognized the factions were a powerful political force that he could harness to his own ambition, and so, while his uncle was campaigning against them, hanging them when necessary, Justinian was cultivating them:

> He appeared to favour one party, namely the Blues, to such an excess, that they slaughtered their opponents at mid-day and in the middle of the city, and, so far from dreading punishment, were even rewarded; so that many persons became murderers from this cause. They were allowed to assault houses, to plunder the valuables they contained, and to compel persons to purchase their own lives; and if any of the authorities endeavoured to check them, he was in danger of his very life: and it actually happened that a person holding the government of the East, having chastised some of the rioters with lashes, was himself scourged in the very centre of the city, and carried about in triumph.[16]

The reasons that the future emperor chose the Blues over the Greens are unknown. Though the composition of the two factions differed in some subtle respects—the Blues were more suburban, more likely to earn their livings from owning land, and more "orthodox," while the Greens were more urban, more likely to be merchants or artisans, and, as with Justinian's disputant, likely to be Monophysite,[17] it seems likely that the selection was arbitrary. If so, it is yet another example of a random choice with a long, and unintended, tail of consequence.

It was, after all, how he met Theodora.

———

Theodora's rise to the apex of the empire is even more improbable than Justinian's. If the best-known story of her childhood is as accurate as it is revealing, her journey began in the same place as Justinian's debate: the Hippodrome.

The great leisure palace was first and foremost a stage for chariot races, but not exclusively, so long as Constantinople's jugglers, acrobats, and animal trainers were available to provide between-race entertainment. One such, Acacius, earned his living as a trainer of bears, who were pitted against mastiffs in one of the most enduring of blood sports. Most of the Hippodrome's entertainers were associated with one of the factions, who after all had started out managing stables for charioteer's horses, and Acacius was no different. He was, in fact, a full-time employee of the Greens, on whom his entire family—wife and three daughters—depended for what was very likely their tenuous hold on a modest prosperity.

When Acacius died, his wife immediately remarried another animal trainer, hoping to retain the same life that Acacius's skill and connections had been providing. Some time around the year 505, she marched onto the floor of the Hippodrome accompanied by her daughters, eight-year-old Theodora, her older sister Comito, and younger sister Anastasia, and presented them before the box containing the leaders of the Green faction, pleading that they grant her new husband her former husband's job.

The Green leadership rejected the plea of Theodora's mother, and with no more thought than a likely desire to provoke their rivals, leaders of the Blues extended an offer of employment to Theodora's new stepfather, unintentionally winning the lifelong loyalty of a small girl.

It would be some years before that loyalty seemed like an asset to the Blues. By the time Theodora was eleven—her actual birthdate is unknown, but is generally thought to be 497—she had followed Comito into the burlesque theaters of Constantinople; by the time she was sixteen, she was working as a prostitute, and had possibly given birth to a son. She definitely bore a daughter before she was eighteen, and shortly thereafter had become both notorious and—not coincidentally—very successful. In her most famous routine, an homage to the legend of Leda and the Swan, Theodora would "spread herself out and lie face upwards on the floor. Servants on whom this task had been imposed would sprinkle barley grains over her private parts, and geese trained for the purpose used to pick them off one by one with their bills and swallow them. Theodora, far from blushing when she stood up again, actually seemed to be proud of this performance."[18] But while salacious stories of Theodora abound—the historian Procopius describes the bearkeeper's daughter as frequently appearing in public virtually naked, having sex with thirty or more men at orgies, and complaining that Nature had constructed her so that she could only have sex via three orifices—the most persuasive portrait is that of a racy comedienne whose interest in a sexually provocative image was largely its proven ability to fill seats.

(That said, while it is true that most of the truly scandalous tales about Theodora's early years are works of political hostility as well as prudishness, they are not uncorroborated. John of Ephesus—a Monophysite bishop and Theodora's friend—calls her "Theodora *ek tou porneiou*" or "Theodora-from-the-brothel.")

In her early twenties, her future on the stage less appealing than her past, Theodora took up with an imperial diplomat named Hecebolus, and accompanied him on his next posting as governor of a North African province. There they fell out, and Theodora made her way to Alexandria, which, with the accession of Justin in 517, had become a haven for Monophysite refugees. In another example of the law of unintended consequences, Theodora, who until then had shown little interest in matters of dogma, fell under the spell of Timothy, the Patriarch of Alexandria, at that time the most prominent Monophysite in the empire. The spell was to last until Timothy's death, by which time Theodora had far eclipsed him in both prominence and influence.

No record exists of her first meeting with Justinian, but, after her return to Constantinople in 522, she was certainly out of the courtesan business,* and it is likely that the two were introduced at some social event sponsored by the Blues, which would likely make it the single most important event in the history of faction politics. However it occurred, it appears to have been a love-at-first-sight thunderbolt. The middle-aged, balding Illyrian and the beautiful twenty-five-year-old

---

* In later tradition, she was living in a modest house making her living spinning wool when they met, a house that would become the church of Saint Pantalaimon after Theodora became empress.

ex-courtesan almost immediately began living together in Justinian's residence at the Palace of Hormisdas on the southern shore of the city, and it is worth remembering that among the tons of invective poured over them in subsequent years by doctrinal opponents and political enemies, not a single word accuses either of betraying the other. As will be seen, their partnership was total, their love for one another complete.

Not surprisingly, the two were eager to legitimize their relationship, but in order to do so, they had to surmount two obstacles, one legal, and the other personal. In fact, Justinian's first legislative achievement was supervising the draft of a law that not merely permitted a "penitent actress [to] apply for an imperial grant of marriage" but, in the case of a person raised to patrician status, eradicated *ex post facto* any "other blemish."[19] But, while Justinian was able to convince his uncle to rescind the law that prohibited marriage between high officials and actresses, or even ex-actresses, he was thwarted by the objections of Justin's wife. The onetime slave girl, who had changed her name from Lupicina to the more aristocratic-sounding Euphemia, was disinclined to be flexible on any matter that might undermine her own tenuous hold on respectability. Only with her death in 524, and the passage of the new law in 525, were Justinian and Theodora permitted to enter into the most successful marriage in the history of statecraft. Countless decrees from Justinian's reign refer to Theodora as "our most pious consort given us by God" . . . love letters in the form of imperial legislation.

In the spring of 527, Justin, who had descended into senility years earlier, died. On April 4, 527, Theodora and Justinian were crowned emperor and empress by the patriarch, Epiphanius, in the great church built by Theodosius II. Immediately following the coronation, they led a procession on a celebratory march that ended in the imperial loge of the Hippodrome, the Kathisma, there to greet their subjects. If anyone noted the contrast between the walk from church to arena and Theodora's journey from the floor of the Hippodrome to the Kathisma, history has failed to record it.

———

Though Theodora had taken on a more modest way of life since her days on the stage, she hadn't completely foresworn public display. The empress cheerfully accepted many of the public appearances required of the sovereign, appearances that her introverted husband was happy to forego. The result was to reinforce the emperor's isolation in his capital, which made him extraordinarily dependent upon the men he selected to act in his name. Luckily, he was extraordinarily good in his choices.

It is unlikely that Justinian learned the leadership skill for which he is most lauded while immersing himself in Christian doctrine, and by the time he was serving as Justin's regent, he had already mastered it. By default, the likeliest assumption is that Justinian was born with an instinct for selecting men of ability to do his bidding, since despite the emperor's training in—and, to be fair, love of—the arts of rhetoric, refined speech and writing, and the arcana of religious dogma, his lieutenants exhibited few if any of these skills. Mastery of prose style and possession of orthodox views were, to Justinian, no barrier to promotion, but counted for nothing compared to intelligence and determination. Virtuosity mattered far more than virtue; a good family was the sort that gave its sons native ability rather than noble ancestors. Above all, the emperor's men possessed ambition enough to be successful, but enough loyalty to value their emperor's success above their own; Justinian learned early that he could more easily bind a wealthy man to his will than a poor one, provided that he was the source of his wealth.[20] If he didn't excite the adoration of his lieutenants like an Alexander or a Caesar, he was also never betrayed by any of them.

Which isn't to say that they never betrayed one another. While Justinian may not have encouraged loathing among his closest advisers, he was clearly unbothered by it. History records no hatred better nurtured than that exhibited by the three who occupied the inner circle of Justinian's court—by Theodora for her husband's de facto Prime Minister, John the Cappadocian, and by the Cappadocian for Justinian's greatest general, Belisarius.

According to Procopius, John was "a man of the greatest daring and the cleverest man of his time."[21] He was also crude, unlettered, hardworking, highly efficient, and personally corrupt though institutionally incorruptible; that is to say, he would not hestitate to enrich himself, but could not be paid off to, for example, alter tax or fiscal policy. During the first two years of Justinian's reign, the Cappadocian, who had joined the imperial service years earlier as a clerk, had accumulated some of the highest titular honors available, from *illustris* to head of Justinian's First Law Commission to Praetorian Prefect: effectively, Justinian's chief minister. In all cases, his real job was the administration and—especially—the financing of the empire. Consider, therefore, that despite the fact that his contemporaries accused him of every manner of debauchery, greed, sadism, and insatiability, John really did reform huge areas of imperial activity, and not always in ways to his own advantage either in wealth or power. Tax collectors— and John was nothing if not enthusiastic about taxes—are rarely well-liked; the popular name given to one of his subordinates, Alexander the Scissors,[22] suggests hostility at the very least. But while John may have been a boor, a glutton, and a bully, it seems likely that he accumulated enemies as much because of his institutional efficiency as his personal abrasiveness. Beneficiaries of John's administrative skill were more likely to be working-class provincials, while those who were impoverished by him were not; by definition, one needs some wealth in order to be made poor. Thus, the record of the period is somewhat distorted by the fact that it is overwhelmingly written by John's victims. Families who had, because of their high status, long escaped taxes, escaped no more. Imperial employees, who were either feather-bedding or simply employed in areas that the emperor was no longer willing to support, were fired.[23] Many of them headed for the imperial capital, where, reinforcing one another's resentments, they turned John into the monster of popular legend.

Popular legend, on the other hand, has never been anything but kind to John's great rival, the general Belisarius. Born sometime around 505 in what is today western Bulgaria, Belisarius—the name is Thracian, and his early service as an officer in a cavalry regiment suggests that his family were minor nobles—was the most extraordinary of all "the incredibly brilliant team of generals that served [Justinian]."[24]

When Edward Gibbon compared Belisarius to Alexander the Great, he was simply another* in a long line of admirers of a soldier whose life had become a cautionary tale of virtuous though unrecognized behavior. By the twentieth century, the general's standing as a strategic genius had overshadowed his reputation for virtue; along with Alexander, Hannibal, Marlborough, and Napoleon he is one of the exemplars of Basil Liddell Hart's classic *Strategy*.

When Justinian tapped him for glory, he was both young and a provincial, though the latter was obviously no barrier to success in Rome's army; generals on the army list during Justinian's reign included the Gepid prince Mundus, the Slav Chilbudius, the Armenians Isaac Kamsarakan and Sittas (who married Theodora's older sister Comito), and the Ostrogoth Bessas.[25] When the Thracian was sent east to Mesopotamia, he was an unknown twenty-five-year-old former bodyguard of the emperor; when he left it, a year later, he was well on his way to becoming a legend.

––––––

If you journey northwest from Iraq into the present-day Turkish province of Mardin, you'll be traveling along the path still taken by Kurdish free traders—smugglers, if you prefer—who were making the journey for a thousand years before German engineers began building the Istanbul–Baghdad railway as the first leg of a line intended to run all the way to Berlin. Forty miles southwest from the provincial capital, on the Turkish-Syrian border, is the town of Nusaybin, located at the mouth of a narrow canyon where the Gorgonbonizra River drains out of the mountains of Asia Minor. That canyon, the only route through the mountains for a hundred miles in either direction, controls access to the fertile plains of northern Syria. The strategic importance of the pass had been acknowledged for centuries, partly for reasons of military advantage, partly commercial paranoia. Fearing

* Jean-François Marmontel's 1767 novel *Belisarius* features the title character delivering long speeches against religious intolerance, the nobility, nepotism, and a host of other evils that the author saw infecting his nation; and when the sculptor and painter David applied for admission to the French Academy in 1781, his chosen subject was Justinian's general. In 1937, Robert Graves made him the object of his hero-worshipping novel, *Count Belisarius*, and he is featured in at least two separate twenty-first-century science fiction series.

that commerce would serve as a cover for espionage, Diocletian had established Nisibis, as Nusaybin was then known, as the exclusive entrepôt in which Persian merchants were allowed to trade, and even when its ownership passed to the Persians by treaty, and they turned it into a fortress, the town's status remained unchanged.[26]

Less than ten miles away are the ruins of what was once the Roman fortress town of Dara, built by Emperor Anastasius in 506, as a tripwire that would give Constantinople warning of any Persian military ambitions.

It was a good place for it. For two hundred years, the canyon marked the farthest extent of the Roman and Persian empires, and the provocation of building two fortresses within it insured that it would become a battleground. Nisibis had been besieged twice before—in 337 and 350. Now it was Dara's turn. For the previous two years, Justinian had been engaged in securing—though not expanding—his eastern border. On the diplomatic front, he sent an embassy, headed by Anastasius's nephew Hypatius, to Kobad, while bribing Caucasian clans to rebel against their Persian overlords, and attacking Persian strong points on the borders of the Black Sea. The result of all this activity was an armistice signed in fall of 528. By spring of 529, however, Persian-sponsored Arab tribes were raiding into Syria, and reports of atrocities forced Justinian to respond. The response was Belisarius.

The battle for Dara did not occur at an auspicious time for Rome. In fact, it would have been difficult to pick a time that *would* have been auspicious, since for nearly a thousand years the only armies to regularly best the legions of Rome were in the east. From the defeat of Crassus by the Parthians in 53 B.C.E. to the capture of the emperor Valerian in 259 C.E. by Shapur I, Persian armies had marked the limit of Roman military triumph. More recently, Justinian had sent armies under the Thracian generals Boutzes and Couzes into Asia Minor, where they were soundly whipped; their replacement, the Greek general Pompeius, fared no better. So when Belisarius, the former junior cavalry officer and imperial bodyguard, was appointed commander, not just of Dara, but of all Roman forces in Mesopotamia, a casual observer would be forgiven for thinking it an act of desperation, rather than strategic brilliance. While Justinian had sound intelligence that the Persians had mobilized a formidable army intended to throw the Romans out of Dara for good, his options for countering them were limited. But a welcome needed to be prepared for the Persians' arrival, and Belisarius was the best greeting Justinian could offer.

———

By June of 530, the Persians had arrived. Arrayed against Belisarius was a host of more than forty thousand soldiers, including at least five thousand "Immortals," the elite heavy cavalry that had been the rock of Persia's armed forces since the days of Cyrus the Great. Expecting to find the Romans sheltered behind the walls of Dara, the Persians, under their commander Firuz, also included a train of artillery, sappers, engineers, and the other impedimenta of a planned siege. The 25,000 troops that awaited them were a Roman army. But they were not Roman legionnaires.

The third century restructuring of imperial defense had done far more to change the army's strategic organization than its tactics. Through the third and fourth centuries, heavily armored infantry remained the predominant arm of the Roman military, though the firepower of the legions, as they were still known, had been significantly increased by the addition of archers and slingers. Though foot soldiers are the most conservative of men, the legionnaires' equipment did change with a view to meeting the challenges of war with barbarians who had themselves changed over the centuries. The Roman legion had adopted chain mail in imitation of the Gauls whom Julius Caesar had defeated, and the *gladius,* or short sword, so deadly in close combat, from the Iberians and Ibero-Celts whom they had fought in the Punic Wars.[27] By the fourth century, the gladius had been effectively replaced by the *spatha,* a longer, slashing sword more effective against unarmored and mounted combatants.

The most effective counter to mounted opponents, however, was more cavalry, and throughout the fourth century, Rome had actively recruited mercenaries and allies skilled at combat on horseback, including Syrian Arabs, African Moors, Huns, and Herulians—raiders originally from Scandinavia, a people Procopius called "the basest of men and utterly abandoned rascals"*[28]—and converted thousands of Illyrian and Thracian foot soldiers into cavalrymen. But at the core of this New Model cavalry were units consisting of private retainers of senior officers, sometimes called the *comitatus*, after the similarly named household troops of the emperor.

At Dara, Belisarius numbered 1,500 men in his *comitatus*, and their introduction marks not only a uniquely effective weapon of the imperial army but also a preview of the composition of European armies throughout the middle ages. A modern historian credits it as "Belisarius' inspired innovation . . . an armored rider mounted on an armored horse . . . Unbeatable on the field of battle, but . . . expensive, cruel, overbearing, even mutinous, and owned by their commander."[29] One might well be describing the knights that would dominate the battlefields of Europe from Roncesvalles to Naseby.

By the time that Belisarius's troops mustered at Dara, the transformation of the Roman army from the disciplined squares of infantry, armed with javelins and short swords, into squadrons of heavy cavalry, a transformation begun by Aurelian and continued by his successors, was complete. The most valuable sixth-century Roman soldiers sat astride horses, armed with lances and—far more deadly—the compound bow that was the great contribution of the steppe peoples to missile weaponry. Short enough to be shot on horseback, recurved in a double-S and reinforced with horn and sinew, this tool of battle could send an iron-tipped arrow through chain mail at a distance of 100 yards. A contemporary description of Belisarius's cavalry was given by Procopius:

> [Our] archers are mounted on horses, which they manage with admirable skill; their head and shoulders are protected by a casque or buckler; they wear greaves of iron on their legs and their bodies are guarded by a coat of mail. On their right side hangs a quiver, a sword

---

* This somewhat prudish chronicler was also moved to observe that "they mate in an unholy manner, including men with asses." The Herulians appear as exotics in Procopius's history, with an ancestral home in the quasi-legendary Ultima Thule, the northern land where the sun never sets.

on their left, and their hand is accustomed to wield a lance or javelin in closer combat. Their bows are strong and weighty; they shoot in every possible direction, advancing, retreating, to the front, to the rear, or to either flank; and as they are taught to draw the bowstring not to the breast, but to the right ear, firm indeed must be the armor that can resist the rapid violence of their shaft.[30]

The amount of space that the historian devoted to describing the importance of archery in the armies of Belisarius—Procopius, the general's legal secretary, hadn't written three pages of the first book of his *History of the Wars* before leaping to the defense of the imperial mounted archers against the accusation that they were in any way like the bowmen Homer ridiculed for their cowardice—illustrates the regard that sixth-century generals had for missile weaponry wielded by cavalry.

Procopius's defense of Roman archery, which seems so curiously excessive, marks its author's obedience to two literary traditions, one Homeric, the other (for lack of a better word) Thucydidean. On the one hand, Procopius's dramatic chronicles of the battles that punctuate his history (more accurately, the battles that comprise his history, which is, after all, entitled *The Wars*) are decorated with deeds of literally unbelievable heroism—arrows plucked out of the air, dozens of opponents slain in single combat. The comic-book hero scenes, however, are tied together by accounts that are not only realistic, sometimes startlingly so, but also illuminating, particularly as to terrain, chronology, tactics, and equipment. His writings on Dara, where he had just joined Belisarius as his legal secretary (a sign, perhaps, of the importance both sides gave to diplomacy as well as combat) are a revealing and illustrative eyewitness narrative.

Thus, we know from Procopius that Rome's armored cavalrymen carried shields, darts, lances, and swords as well as their bows, but also that while they were individually formidable, they had yet to show anything like the discipline of their Persian opponents, which went a long way toward explaining their earlier setbacks. An ability to teach and impart such discipline was one of the great gifts Belisarius brought to the eastern frontier.

A gambler's instinct for calculated risk was another. When imperial scouts first brought Belisarius word of the size and nature of the approaching army, he put that instinct to the test: He immediately decided to leave the protection of Dara's walls and fight a numerically superior enemy in the open field. The reception he prepared for the attacking Persians revealed a third gift: a flair for innovation that, by itself, probably justified Justinian's choice. Procopius described it this way:

> Not far from the gate which lies opposite the city of Nisibis, about a stone's throw away, they dug a deep trench with many passages across it. Now this trench was not dug in a straight line, but in the following manner. In the middle there was a rather short straight portion, and at either end of this there were dug two cross trenches at right angles to the first; and starting from the extremities of the two cross trenches, they continued two straight trenches in the original direction to a very great distance.[31]

Behind these earthworks, Belisarius placed a wing of cavalry on either flank; in the center, also behind the trench, stood the young commander and the Roman infantry. But in front of it, he placed the four squadrons of his Hun cavalry. In brief, he had taken the traditional geometry of the battlefield—infantry in front, cavalry in the rear as a mobile reserve—and turned it inside out.

Then, he waited. On the first day after he arrived in June 530, Firuz began the battle, but with a pen. Under a flag of truce, the veteran Persian commander sent a taunting note to his young adversary, ordering up a bath and a meal in Dara for the next day. Despite this seeming confidence, Firuz seems to have been nonplussed by Belisarius's nontraditional defense. Or perhaps he was just a careful commander, as mindful of his quartermasters as his cavalry. Whatever the reason, Firuz spent the first day with his senior officers planning an attack.

While the Persian commander was occupied in the prosaic business of modern warfare, making certain, for example, that his archers were sufficiently well supplied with arrows, one of his officers reverted to the Bronze Age heroics of the *Iliad*, and rode back and forth along the Roman lines daring anyone to meet him in single combat. Procopius describes his general emerging from the Roman command tent to see the challenge accepted . . . but not by a soldier. Instead, "a certain Andreas . . . a trainer of youths in charge of a wrestling school in Byzantium"—a sort of personal trainer to one of Belisarius's officers— accepted the Persian's challenge. Since we are told he was "not one who ever practiced at all the business of war," beginner's luck must have been shining on him, for he knocked the Persian from his horse with a single well-placed blow of his eight-foot lance, leaped to the ground, and administered the coup de grace with a knife.

Belisarius himself applauded, striking the flat of his sword on the boss of his shield, and his troops raucously followed suit. The Persians were "deeply vexed." Their second champion, brandishing a whip, summoned another volunteer. He, too, got Andreas.

Stirrups were still decades in the future of both Persian and Roman cavalry, and the lances carried by the combatants were not, therefore, used in the manner of a late medieval joust; without the stability of stirrups, much lighter lances were carried just above the shoulder, rather than under the arm. Nonetheless, Andreas's second contest combat was still essentially a collision of two armored men on ar- mored horses, bearing lances, swords, and a surplus of hostile intent, the rough equivalent of a motorcycle crashing into a concrete abut- ment at thirty-five miles an hour, and with a similar result: Both men went flying. Without stirrups, neither was able to remount without aid; thus, the fastest man to his feet would have an impressive advan- tage. This time, calling more on his wrestling experience than luck, Andreas quickly gained his footing, sword in hand, while his oppo- nent was still rising to his knee, which was as far as he got.

The morale boost of Andreas's victories lasted until the morning of the second day, when ten thousand Persian reinforcements arrived from Nisibis, giving Firuz a two-to-one numerical advantage. Belisar- ius continued the epistolary phase of the battle with a letter reminding Firuz not to "war without justification." Within an hour, he had a re- ply: "I should have been persuaded by what you write, and should have done what you demand, were the letter not, as it happens, from Romans." Belisarius, having observed the forms, wrote once more to his opponent, informing him that the Romans had arrayed for battle by fastening the insulting letters to their banners.

The message could not have been clearer: "The army of Rome awaits."

———

On the third day, the battle for Dara began in earnest. The Persians attacked on the Roman left, advancing knee to knee under flight after flight of arrows; by the time they were within a hundred yards of the Romans, the archers had taken their toll, and the Persian cavalry proceeded to canter and then trot their horses into position for a line-to-line attack with lances. Under such pressure, the Romans withdrew so rapidly that the Persians became overeager and started to gallop in pursuit. As is nearly always the way with cavalry, what starts as a cohesive attack while horses are walking becomes a melee once they start to gallop. Faster horses quickly outdistance the slower, and a tight formation dissolves. Nonetheless, the Persians rightly believed that if they could force a collapse on the Roman left, their superior forces could easily cross Belisarius's trench, wheel left, and roll up the remainder of the defenders. Firuz might already have been planning the menu for his promised lunch.

What he did not know is that Belisarius had placed his small but fearsome division of Herulian horsemen on the reverse slope of the small hill that anchored the Romans' left flank—available, but hidden from the enemy. The six hundred horsemen rushed around the hill just as the Persians started their own gallop, catching them in their completely unprotected right flank. The Herulians were trained to shoot arrows while moving in the saddle, and probably had time for two volleys before striking the Persians with sword and lance.

Conservative estimates place the Persian casualties at some two thousand in less than twenty minutes, victims of the unforgiving geometry of the battlefield. Because of the limitations of anatomy, humans are evolved to act effectively only in the direction that evolution has pointed eyes and hands. The consequences of this simple fact for military tactics, from Caesar to Napoleon to Patton, are always the same: Troops are more vulnerable on either side than they are in their front, and terribly so in their rear. Virtually the entire library of tactics, as set down in classics from Sun Tzu to Liddell Hart, consists of ornate descriptions of the best way to apply force—clubs, arrows, or .50 caliber machine-gun bullets—from *your* front to your enemies' flank. And, obedient to the Golden Rule of Soldiering, to do so to him before he does so to you. Valens had paid the price for ignoring this rule at Adrianople; the first Persian attackers at Dara had been reminded of it at the cost of a significant part of their front-line troops. The second wave was to pay an even dearer price.

As the first Persian attack was cut apart by Pharas's Herulians, the second pressed strongly against the outnumbered Roman right. Seeing this, Belisarius ordered the four squadrons of Hun cavalry holding the Roman center to prepare for a counterattack. The Huns were more than ready.

For 150 years the world had hated and feared the Hun cavalry, whose ferocity and skill had remade the map of Europe. With the death of Attila, their dominance had been ended by their lack of discipline. Even so, for decades after their armies had been broken, the amazing accuracy of Hun mounted bowmen, and the wildness of their charges, remained the stuff that soldiers used to frighten one another around campfires. Now the Persians were to meet them while they were under the command of a general who knew how to use them.

Once again, the Persians outran themselves as the Romans pulled back over the trench, and when a gap opened between the nearly ten thousand troops that Firuz committed to attack and the remainder of the Persian army, Belisarius gave the order, and the Huns pivoted like a great swinging door. The two closest attacked the Persian right flank; the two farthest, racing their horses at top speed, made it into their rear. With arrows firing at nearly point blank range from two sides every ten seconds, and the Roman front now a wall of shields and spears, the Persian soldiers could only stand and die.

By the end of the day, at least five thousand Persian corpses littered the battlefield, to be collected for burial by Firuz's crippled army and carried back to Nisibis. Only the Romans bathed that evening in Dara, which had been preserved as a strategic anchor for the Roman Empire's eastern flank. The Romans had proved the value of their heavy cavalry.

And Justinian had found himself a general.

PART II

GLORY

CHAPTER FOUR

"Solomon, I Have Outdone Thee"

530–537

> So Solomon overlaid the house within with pure gold: and he made a partition by the chains of gold before the oracle; and he overlaid it with gold. And the whole house he overlaid with gold, until he had finished all the house; also the whole altar that was by the oracle he overlaid with gold. . . . Thus spake Solomon, "The Lord said that he should dwell in the thick darkness. I have surely built thee a house to dwell in, a settled place for thee to abide in for ever."
>
> —1 Kings, 6:21–22, 6:12–13

BELISARIUS MAY HAVE been the greatest Roman general since Julius Caesar—an achievement that is even more notable than it sounds, since generalship was one of the few Roman attainments about which no argument exists. In fact, when compared to the brilliant originality in poetry, politics, experimental science, medicine, logic, drama, and philosophy of their Greek predecessors, the greatest of Roman attainments can seem modest indeed. The efflorescence sometimes called "the Greek miracle" appears to cast a shadow on Rome like a great forest blocking smaller trees from the sun, with Homer, Plato, Aristotle, Sophocles, Herodotus, Socrates, Euclid, and Thucydides towering over Virgil, Marcus Aurelius, and Plutarch.

To be sure, the world's debt to Rome is far more than merely, as has often been observed, the arch and the law. But these are not small things, and an argument could be made that Rome's contributions have benefited substantially more people, with enormously greater frequency, than all of Greek drama and philosophy combined. The value of the inheritance left by Roman engineers and architects alone,* compounded for more than two millennia, is literally incalculable, most especially as the foundation for the building of modern Europe. By the time of Justinian, that legacy had been undergoing constant refinement for five hundred years, and it is therefore somewhat surprising to discover that its apotheosis was built almost literally on the ashes of the greatest civil insurrection in the history of Constantinople.

————

The weeklong riot known to history as the Nika Revolt began and ended at the mile-long Hippodrome that was the emotional heart of Constantine's city. On January 14, 532, the Tuesday morning on which the Nika drama began, the Blues and Greens that dominated the Hippodrome's races and other contests showed up at the stadium in greater than normal numbers, and with emotions running high. The previous Sunday, Eudaemon, Prefect of the city of Constantinople, had sentenced seven violent offenders from both Blue and Green factions to death, four by beheading, and three by hanging. No doubt he was acting on direct instructions from his emperor, who, while still actively favoring the Blues, was by now attempting to apply legal discipline to each gang in a manner that would appear impartial. In practical terms, this meant that justice need not be blind, so long as her scales balanced.[1] In the event, equality is what he got, for the hangman bungled the executions, and one man from each faction fell to the ground alive. The two survivors, aided by monks from the monastery of Saint Conon, sought sanctuary on the other side of the Golden Horn, in the church of Saint Laurentius, where Eudaemon sent guards to ensure that the fugitives would not escape.[2]

Tuesday was the next scheduled race at the Hippodrome, and for once both factions were in complete agreement: What each wanted, more even than advantage over the other, was clemency for the men who had cheated the hangman three days previously. When the crowd's plea for mercy was frustrated, its mood, never particularly peaceful, quickly deteriorated. By the time that the twenty-second

————

* The equally notable bequest left by Roman jurists is described in Chapter 5.

race of the day had been run, the Blues and Greens had agreed on a slogan—*Nika*, meaning "victory" or "conquest"—and a target: the prefect, and the prison that he ruled. That evening the crowd, now a mob, broke into the prison, released all the occupants, and set fire to the Chalke entrance on the north side of Justinian's palace. A wind from the south spread the flames north to the senate building and to the cathedral church, the Hagia Sophia, the second to stand on that site, and the second to be burned down.*

Fire, not riot per se, was the great danger. Under Augustus, Rome had its first regular police force and fire department, but by the fourth century neither of the imperial capitals had preserved his innovation, depending entirely upon volunteer firefighters. Though each of the thirteen regular regions of Constantinople had a nominal police chief, a *curator*, his entire constabulary consisted of one *vernaculus* (a publicly owned slave who acted as the curator's messenger service) and five *vicomagistri*. Much of the history of the city was, not surprisingly, punctuated by riots for which the only solution was the application of direct military force—when, that is, the army could be relied upon.[3] Moreover, every riot carried with it not merely the threat, but the reality, of citywide fire. Nika was only the most recent example.

By Wednesday, despite Justinian's order that the races be resumed, the mob had developed a taste for riot, and they burned a number of buildings on the north side of the Hippodrome, including the enormous Baths of Zeuxippus. But along with the growing anger came a change in its direction. For the ambition of the Blues and Greens—at least temporarily self-named the "Green-blues" or *prasino-venetoi*—had by now swollen well beyond a simple demand for mercy. Feeling their power, the leaders of the revolt demanded the dismissal of Eudaemon; of the Quaestor, a legal historian named Tribonian; and of John of Cappadocia, currently serving as Praetorian Prefect. The grievances were different—the street thugs of the Blues and Greens were hostile to the city's chief constable, while members of the middle and

---

* The first of Constantinople's churches to be named for Holy Wisdom was built by the Emperor Constantius II in 360, and was burned down in an uprising against the emperor Arcadius in 404. The one destroyed by the Nika mob had been rebuilt in 415 by the emperor Theodosius II.

upper classes resented both Tribonian's corruption and the ruthless efficiency of John's tax collectors—but the anger was indistinguishable.

And, while subsequent events show that Justinian was unconcerned about the truth of the accusations, he cared greatly about the fury with which they were delivered. He accordingly acceded to the mob's wishes, appointing a new Prefect of the City, a new Quaestor, and Phocas, "a man of the highest probity,"4 as Praetorian Prefect. Feeding the tiger, however, failed to sate its appetite. No one can know whether the riots began as a genuine expression of popular resentment against corruption. By its second day, however, there is little doubt that the rioters had been coopted by forces that wanted not to reform Justinian's government, but to replace it. So, on Thursday, a group of senators saw the unusual comity between Blues and Greens as the tool to rid themselves of Justinian and place a more deferential ruler on the throne.

Over the next three days, Constantinople was witness to almost constant street fights and arson. The mob tried to burn the Praetorium, the prison from which they released the prisoners on the riot's first day; the wind, again from the south, carried flames north of the burned-out hulk of the Hagia Sophia. The rioters occupied the Octagon, a building due north of the Hippodrome and burned it as they did churches, hospices, and public baths. The riot spread down Constantinople's main east-west thoroughfare, the Mese, toward the forum of Constantine, destroying everything in its path.

By Saturday night, the complex of palace buildings was effectively isolated, and the emperor besieged. The paranoiac distrust that figures prominently in all descriptions of Justinian's personality had never served him better (and never would again): He correctly sensed that the rioters were being directed by members of his own senate, and evicted those senators then occupying the palace with him, notably two nephews of Anastasius: the one-time envoy to Persia, Hypatius, and his brother, Pompeius. Possibly the removal of the former emperor's nephews inspired Justinian's next move; twenty years previously, Anastasius had faced another insurrection, this one in protest of his Monophysitism, and had appeared in the Hippodrome with an offer to resign. The next day, on Sunday, Justinian presented himself before a large portion of the rioting mob at the Hippodrome, and offered them—swearing on a copy of the Gospels—a general amnesty. But what worked for the well-trusted Anastasius failed Justinian. No one believed him.

Meanwhile, the plotters—it was, by this point, a more appropriate description than rioters—were in hot pursuit of a new candidate for the imperial throne. Their first choice was yet another of Anastasius's nephews, Probus, but he had prudently left the city at the start of the troubles. His cousin, Hypatius, either slower or more ambitious, was discovered at home and taken to the Forum of Constantine, there to be crowned with a golden chain over the objections of his wife, Maria, who saw, presciently, that the chain resembled a noose much more than it did a crown.*[5]

Justinian, having failed to defuse the insurrection with oratory, now tried to suborn it with bribery, sending a trusted retainer, an Armenian eunuch and court chamberlain named Narses, to deliver purses filled with gold to faction leaders—and, possibly, to act as an agent provocateur. Had he a week or two to foment the natural distrust between Blues and Greens, it seems likely that the riot would have collapsed on its own. But with tens of thousands of armed men less than half a mile from his palace, a usurper with a strong claim on the throne leading them, and his own palace guards—the same *excubitores* that had once been led by his uncle Justin—sitting out the battle, Justinian's nerve failed, and he announced to his closest advisers his plan to flee to Heraclea in Thrace. The plan was endorsed by his generals and ministers, bodyguards and counselors.

Only Theodora opposed her husband's withdrawal. Her speech deserves quoting at length:

* Hypatius was not precisely enthusiastic about the honor; after he led his new subjects to the Hippodrome, Hypatius secretly sent a message to the emperor, urging an immediate attack on the closely packed mob.

As to the belief that a woman ought not to be daring among men or to assert herself boldly among those who are holding back from fear, I consider that the present crisis most certainly does not permit us to discuss whether the matter should be regarded in this in some other way. . . . My opinion then is that the present time, above all others, is inopportune for flight, even though it bring safety. For . . . one who has been an emperor it is unendurable to be a fugitive. May I never be separated from this purple, and may I not live that day on which those who meet me shall not address me as mistress. If it is your wish to save yourself, O Emperor, there is no difficulty, for we have much money, and there is the sea, here the boats. However consider whether it will not come about after you have been saved that you would gladly exchange that safety for death. For as for myself, I approve a certain ancient saying that royalty is a good burial shroud.[6]

In addition to his wife's steadiness of nerve, Justinian could count on one other bit of good fortune—the recent return to Constantinople of his most loyal and capable general, Belisarius. Between Belisarius's personal retainers, and the Herulians who had likewise accompanied their commander, Mundus, just recalled from the Danube, approximately fifteen hundred soldiers were available for the defense of the throne. Those soldiers had been involved in small skirmishes with rioters for several days before the emperor, braced by his wife's courage, was ready to use them decisively. The following day, Monday,* was, indeed, the day of decision.

Belisarius's original plan was to execute a *coup de main,* a quick strike at Hypatius himself, who was still visible in the Kathisma, the imperial loge high in the southern end of the Hippodrome. A tunnel and winding stair connected the Kathisma directly with Justinian's palace, and was the perfect route for a surprise attack . . . or it would have been had it not been defended by the still-neutral imperial

---

* In truth, there is some doubt whether the final attack by Belisarius and Mundus occurred on Sunday or Monday; Procopius is mute on the subject. But given the number of events that are reliably accounted for on Sunday—Justinian's plea to the Hippodrome crowd, Hypatius's arrival, Narses's mission—it seems impossible that the denouement could have occurred any earlier than Monday.

guards, who were refusing passage to both Belisarius's troops and Hypatius's mob. Instead, Belisarius marched several hundred of his men through the ruins of the burned-out city and around to the opposite side of the Hippodrome, entering at its southeast corner. To his left was a barred door to the Kathisma, defended by Hypatius's supporters; to his right, a mob, many unarmed, but numbering perhaps fifty thousand. Attacking one meant turning his back on the other.

Meanwhile, Mundus had led his Herulians to another of the Hippodrome's entrances, the Nekra gate at the arena's northern end. Knowing this, and calculating that Hypatius's bodyguards would stay by their posts, Belisarius drew his sword and gave the command to charge into the crowd. From the other side of the mob, Mundus did the same.

At the moment that the trap closed, the Hippodrome was so full of rioters that they were physically pressing in on one another, with fifty thousand mostly unarmed people occupying considerably less than 50,000 square meters. (The Hippodrome was 117 meters wide by 500 or so meters long, but the arena floor was smaller in both dimensions.) In close ranks five hundred armed and armored troops would make a line four or five files deep, the entire width of the Hippodrome floor, and Mundus and Belisarius disposed of at least that many men at either end of the arena. The two battalion-sized units needed to traverse only a few hundred meters before they met in the middle of the killing ground, by all accounts looking like nothing so much as a modern threshing machine sweeping through wheat. It was, not surprisingly, a massacre; even given the tendency to exaggeration in the numbers of every ancient battle, the death toll was gigantic. Thirty thousand is Procopius's figure; *all* the other contemporaneous accounts, including those of John Lydus and John Malalas, are higher, ranging as high as fifty thousand. Were these off by an order of magnitude—if, for example, they are something like a casualty figure, with the merely wounded outnumbering the dead ten to one—the body count would still be the equivalent of the September 11 attacks on the World Trade Center, in a city a tenth the size of New York.

Given the scale of the damage that accompanied the insurrection, and of the slaughter that ended it, Justinian's eventual response seems almost measured. He did execute both Hypatius and Pompeius, and banished a number of senators suspected of involvement, but it seems almost as though the emperor believed that, with his throne saved, he had a more urgent duty than revenge—rebuilding his city, and especially its most important church.

———————

The significance of that church to its builders is difficult for a modern visitor to appreciate. Henry Adams argues that the great cathedral-building spree of thirteenth-century France—an "intensity of conviction never again reached by any passion"[7]—was fundamentally an economic enterprise, a nation's treasure invested in assets to be redeemed in the afterlife. Constantine's fusion of empire with Church demanded a similar diversion of capital—witness the Church of the Holy Sepulchre in Jerusalem—but the great churches of late antiquity seem to be rooted more in politics than economics. The Christians of Constantine's New Rome, his city of God-and-Emperor, needed a leader, and the church's leader needed a cathedral—from *cathedra,* originally the name of the seat occupied by a bishop. The first cathedral built by Constantine was the Hagia Irene, north of the Hippodrome. His second son, Constantius, moved the bishop's seat to his new church, the Megale Ecclesia, or "Great Church" in 360. Shortly thereafter, it acquired its new name: Hagia Sophia.

The etymology of the name is almost as ornate as the church itself. Though sometimes mistranslated as "Saint" Sophia, the name actually honors an anonymous biblical text entitled "The Wisdom of Solomon," which attempted to incorporate elements of Greek philosophy into the mystical traditions of Judaism. The text, which never made it into the Jewish canon, is distinguished by its personification of Wisdom, an interlocutor between God and man, named in Greek "Sophia." Thus, Hagia Sophia: Holy Wisdom.

The new Church of Holy Wisdom was damaged only a year after it was built, prompting Constantius's cousin and successor, the pagan Julian the Apostate, to write "See what sort of church the Christians have. If I return there from the Persian war, I shall store hay in the center and turn the aisles into stables for horses."[8] He was whistling past the graveyard of paganism. By 363, Julian was dead, but the Christian character of Constantinople was never healthier. As the fourth century came to a close, the occupant of Constantinople's cathedral was second in religious authority to only one other in the Christian world, and some would say not even to him.

This made for a not always amicable relationship with the emperor, the city's other great authority. When the Emperor Arcadius recruited John Chrysostom to become the new bishop of the imperial see in 398, he was no doubt bargaining that becoming the leading churchman of Constantinople was preferable both to remaining at his home pulpit in Antioch and even moving to the other great eastern city, Alexandria. Arcadius succeeded in placing an irresistible temptation before the cleric, but the bishop proved far more than a trophy for the emperor's vanity; his combination of great eloquence—his surname, *Chrysostomos*, means "golden-mouthed"—and a reformer's passion for the poor at first charmed, and finally discomfited, his wealthy patrons. When Chrysostom fell out with his most prominent supporter, the Empress Eudoxia, his fate was sealed, and on June 24, 404, he was conducted into exile by imperial soldiers. Within weeks, the city was convulsed by battles between supporters and enemies of the rabble-rousing bishop, and the Great Church was burned down. It would be eleven years before a new church was dedicated on the site by Emperor Theodosius II.

Nothing is known definitively about the Theodosian Church, but the best documentary and archeological evidence—chronicles describing the nature of services held there, as well as excavations made by the German A. M. Schneider in the 1930s—suggest that it was built on a rectangular floor plan, with aisles running along the long axis. If so, it was very much in keeping with one of the great architectural traditions of Christianity: the western basilica.

The Roman world, always hungry for monumentally large interior spaces, had developed a new taste for them with the ascendancy of Christianity. Christian worship services were both frequent and, once Christianity became the official religion of the empire upon the death of Constantine in 337, virtually universal. By the end of the fourth century, churches had tended to adopt one of two architectural models. Those in the west were built in the shape of basilicas (from *stoa basileos,* or king's room) and were the first really large buildings for Christian worship. The typical western basilica was built around five (sometimes three) longitudinal aisles, the middle one—the nave—higher than the surrounding four, with a semicircular blister—the apse—on the end opposite the entrance, with a half-dome vault above. Constantine built some of the world's most notable basilicas during the first decades of the fourth century, including the original St. Peter's and Lateran in Rome and the Church of the Holy Sepulchre in Jerusalem, and the form retains its popularity to the present day. Partly this is simple liking for tradition, but mostly it is because the basilica form possesses "the triple merits of being easy to build, free of undesirable pagan connotations . . . and almost ideally fitted for the accommodation of large congregations attending a priestly liturgy."[9]

Those churches that were not rectangular basilicas were typically built in the round, though examples of circular or octagonal structures in the grand size are fewer; the most prominent fourth-century versions are probably the Church of San Lorenzo in Milan, and the Golden Octagon in Antioch, described by the early Christian historian Eusebius as "a church unique in size and beauty . . . raised inside to a great height in the form of an octagon surrounded on all sides by two story spaces. . . ."[10] But the relative paucity of centrally planned *churches* does not mean that the world of late antiquity lacked for models of monumental circular *spaces.* Nero's Golden House, the Domus Aurea; the baths of Agrippa; and, biggest of all, the Pantheon of Hadrian, are among the dozens of great domed spaces familiar to the Mediterranean world.

Debate continues to this day over the reasons for choosing one form over another in early church design. Some architectural historians hold that the first designs for churches were arbitrary, with builders simply using the most familiar forms they knew. Others suggest more deliberation, arguing that the early Christian basilica was designed to awe lay worshippers congregated in the nave by alternating light and shadow: A well-lit atrium giving way to a dark narthex and then to a bright nave—a "commodious ordering of the space"[11] both in height and depth, with the aisles giving the impression of infinite length. Similar is the belief that the form of the centrally planned churches of the east was determined by the Greek liturgy, which frequently demanded a nave occupied only by clergy with lay congregants gathered in the surrounding aisles. Though the architects of late antiquity were working fifteen centuries before Walter Gropius articulated the design principle that "form follows function," they would have understood it in their bones.

Whatever the origins of the two architectural themes, by the early sixth century they had come to symbolize the widening gulf between the orthodox west and the Monophysite east. When Justinian took the throne, both types of church structure were found throughout Constantinople—the basilicas usually with three aisles, the centrally planned churches octagonal, square, or round—though the theological and esthetic gulf separating them made for an uneasy coexistence, like a Quaker meeting house built next door to the Mormon Tabernacle. Partly, this was because of disparities in size. Constantinople's circular churches tended to be less than thirty feet across, while one of its grandest basilicas, St. Polyeuctos, a private chapel built in 527 by the wealthy noblewoman Anicia Juliana,* rivaled the cathedral in size.[12]

The rioters who burned down Theodosius's church had presented Justinian with more than simply an opportunity to build a monument to his reign, and more even than a chance to honor God. It was an occasion to heal a breach, and he would not let it pass. It would not be enough, therefore, to simply build a larger version of St. Polyeuctos, or the Theodosian cathedral. The new church needed to be different in kind, as well. It would need to suit both eastern Christians,

---

* The builder of St. Polyeuctos, from the same Anicii family as Boethius, was a descendant of Theodosius and Constantine, and her son was married to Anastasius's daughter. Whether Justinian saw her as a rival for his throne is unknown, but one can't avoid thinking that he regarded her building as an accomplishment to be bettered.

whether Monophysite or orthodox, as well as westerners. From a purely practical standpoint, it must combine the liturgical need for regular processions down a long nave with the political advantage of permitting as many congregants as possible to see the emperor himself at worship. It needed to be beautiful, and to inspire literal awe. It needed light, brilliant light, no less than the gothic architects, who, in the words of Henry Adams, "needed light and always more light, until they sacrificed safety and common sense in trying to get it."[13]

The architects chosen to meet these needs were both members of the Greek-speaking elite that dominated the academies and professions of the eastern Mediterranean throughout late antiquity: Anthemius of Tralles and Isidore of Miletus. Anthemius's family, from Lydia (near Izmir, in modern Turkey) was especially prominent, with one brother, Olympicos, a well-known jurist; and his father, Stephanus, and brothers Dioscorus and Alexander, all physicians.* At the time they received the commission to rebuild the Hagia Sophia, neither Anthemius nor Isidore could accurately be called an experienced architect; both Agathias and Procopius refer to them not as architects, but *mechanikoi,* which translates more precisely as "engineers," though even that term is inaccurate.[14] The fourth-century mathematician Pappus of Alexandria wrote that the "science of mechanics is held by philosophers to be worthy of the highest esteem, and is zealously studied by mathematicians, because it takes almost first place in dealing with the nature of the material elements of the universe. . . . The ancients describe [*mechanikoi*] as 'wonder-workers.' "[15] In truth, both Anthemius and Isidore were academic mathematicians, teaching geometry in Antioch before they received the summons from Justinian.

As mathematicians, however, their reputations were considerable indeed; Isidore wrote a commentary on Heron of Alexandria's geometry text, *On Vaulting,* and is popularly believed to be the author of the fifteenth and final book of Euclid's *Elements,* while Anthemius is independently famous for his seminal work on the geometry of parabolas

---

* According to the historian Agathias, the *real* success story of the family was yet another brother, the grammarian Metrodorus.

and reflection ... so much so that generations of mathematicians would use Anthemius's book *On Burning Mirrors* in their attempts to replicate the mysterious mirrors supposedly invented by Archimedes as a weapon.

(Mirrors fascinated Anthemius his entire life. Years after the completion of the Hagia Sophia, the architect—now a wealthy and famous citizen of Constantinople—was the loser in a lawsuit brought by his upstairs neighbor, the equally famous orator Zeno. He took his revenge like a proper engineer, first simulating an earthquake with a steam line that he surreptitiously ran into Zeno's apartment, then exploding noisemakers to mimic the sound of a thunderstorm. Finally, he put his geometric talents to practical use—or, at least, practical-joke use—employing a pivoting parabolic reflector to shine light at all hours into Zeno's sleeping chamber. When Zeno asked Justinian to intervene, the emperor declined to punish his architect, writing that even he "cannot intervene against Zeus the Thunderer and Poseidon the Earth-Shaker."[16])

The early stages of the design of the Hagia Sophia are, unfortunately, less well documented than its architects' juvenile sense of humor. A story has it that Justinian dreamed that he saw a wise man bearing a large silver plate containing an engraving of the church that would rise on the ashes of the old Great Church. In the dream, Justinian told the wise man "If I had this plate, I would use it as a model for my church."[17]

It is likely that Anthemius and Isidore began with a less fanciful version of the client's original commission: Expanding the space of the Theodosian church while preserving sight lines for the maximum number of congregants to be able to watch the emperor at prayer.[18] The geometric form that places the most area at the shortest distance from a single point is the circle, but since the need for processional (and, possibly, a resistance to the "pagan" associations of circular spaces) precluded a purely octagonal or circular nave, the architects proposed squaring off the nave at the end opposite the entrance and adding two semicircular areas, called hemicycles, to each side. The new church would increase the length of the nave of the Theodosian church while widening the worship area, and so increase the scale of the entire structure. It would enable the clergy and emperor to lead dignified processions down the center of the church, and present the best view of their prayers to the largest possible audience. And, not at all coincidentally, Anthemius and Isidore's floor plan would harmonize patriarch with ruler, and join east to west every time congregants walked from one end of the church to the other.

But any building, and particularly a church, is more than just where its occupants walk. It is also what they see. For fifteen centuries, the marble and granite of the Hagia Sophia have awed visitors,* who looked at hard stone and saw instead "a meadow with its flowers in full bloom . . . the purple of some, the green tint of others, and those on which the crimson glows and those from which the white flashes, and . . . those which Nature, like some painter, varies with the most contrasting colors."[19] Others were stunned by the mosaic work, now destroyed, but once "so fine that no brush could better it."[20] Ibn Battuta, the indefatigable Arab traveler of the thirteenth century, saw the structure only from the outside, and nonetheless was struck by the beauty of the marble enclosure around the church. In the end, however, the glory of the Hagia Sophia was always in its design more than in its decoration: In the finished structure, "space, form, and light spoke for themselves."[21]

Even more astonishing than the beauty of the church's design is the speed with which it was produced. The two architects could not have met with their imperial client much less than two weeks after the flames of the Theodosian church had been extinguished. There is little doubt that the final design for the Hagia Sophia was as much a product of the trial and error of the building process as of the sketches produced by its architects. But the church's rough outlines—nave, gallery, and dome, "space, form, and light"—were in hand before the first workmen started digging the foundation, on February 23, 532, less than six weeks after the January riots.

---

* Most of them, anyway. Mark Twain, in *Innocents Abroad*, called it "the rustiest barn in heathendom" but was never one to let awe get in the way of a good punch line.

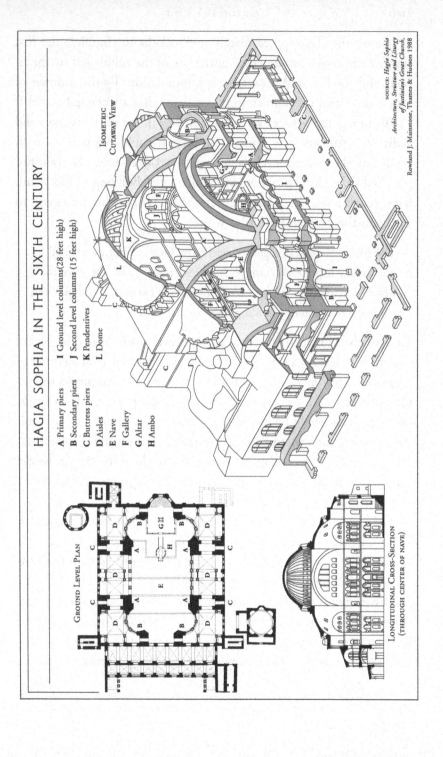

HAGIA SOPHIA IN THE SIXTH CENTURY

A Primary piers
B Secondary piers
C Buttress piers
D Aisles
E Nave
F Gallery
G Altar
H Ambo

I Ground level columns (28 feet high)
J Second level columns (15 feet high)
K Pendentives
L Dome

ISOMETRIC CUTAWAY VIEW

GROUND LEVEL PLAN

LONGITUDINAL CROSS-SECTION
(THROUGH CENTER OF NAVE)

SOURCE: Hagia Sophia
Architecture, Structure and Liturgy
of Justinian's Great Church,
Rowland J. Mainstone, Thames & Hudson 1988

It took the crew approximately five months to complete the first stage of the construction, clearing the site of the rubble left by the rioters and the fire, and excavating the foundation. By the summer of 532, Isidore could announce to his emperor that a rectangular trench 234 feet long and 195 feet wide marked the outline of the corners and walls of Justinian's new church, and work could begin in earnest, pounding the ground-level piers twenty feet deep into Constantinople's bedrock. Those piers, on which the entire building would stand, have become legendary for their strength. Writing his panegyric to Justinian-the-Builder, Procopius wrote:

> The [piers] were held together neither by lime . . . not by asphalt . . . nor by any other such thing, but by lead poured into the interstices, which flowed about everywhere in the spaces between the stones and hardened in the joints, binding them to each other.[22]

Each of the four main piers, at ground level made of granite blocks one to three feet on a side, was placed inside the trench, at a corner of a square 118 feet on a side.* Outside the main piers, forming a rectangle 164 feet by 118 feet, were the four buttress piers, made of brick, each of which would be joined to the main piers in order to transmit the weight of the structure to an even more massive, and therefore stable, base. A set of secondary piers, at the outside of the trench, would support the dome and the vaults over the church proper. All would be held together by a series of arches.

The arch is one of those brilliant innovations that, once discovered, leaves civilizations figuratively scratching their heads and saying, "Why didn't *we* think of that?" Spanning a given space with horizontal beams is a losing game: The longer they get, the more stress they must withstand, and even a granite beam cannot span a space longer than about sixteen feet without cracking. By converting all the stress that fractures the middle of those stone beams—technically *tension*—into *compression* on stone piers, larger and larger spaces could be

---

* Contemporaneous measurements are given in the "feet" used in Constantinople during late antiquity, which is a bit larger than the modern version; all dimensions have been reduced accordingly.

spanned. To be sure, all that compression isn't exactly straight. Much of it hits at an angle, which tends to push the base over. Like a pillar built of books, it is quite stable so long as the only stress is straight down through the center of the book on top. But shift the pressure even slightly off center, and the pillar is likely to collapse. As a result, the books receiving all that stress—the piers—had to be either truly massive (so that the stress is countered by sheer weight), or buttressed in some way.*

From the arch to the dome is a far smaller leap than from the post-and-beam to the arch; a dome, after all, is essentially a group of arches set in a circle. In their early incarnations, the limitation on both arch and dome was the ability of craftsmen to shape the stones carefully enough to create blocks in precisely the wedge shapes needed for a particular arch. Despite their mathematical sophistication in most other respects, the architects of antiquity lacked a proper geometric solution to the ideal form of the arch. (It was not until 1675 that the English polymath Robert Hooke described mathematically the shape of an arch loaded in pure compression, that is, with no tension, by showing how it describes an upside-down version of the catenary curve of a hanging chain.) As a result, the only way they could design an arch, and its component stones, was completely by eye, and those eyes and the hands that could cut stones to such tolerances commanded high prices. Rome overcame this drawback with typical ingenuity, first replacing stone with bricks and mortar . . . and expensive stonecutters with relatively cheap bricklayers.† Both solutions would feature prominently in rebuilding Constantinople's cathedral; the entire structure of Anthemius and Isidore's church would be made up of stone, brick, and mortar.

---

* Or, as is the case with those familiar Roman aqueducts, built in a series so that each arch transmits the angular stress to the arch next door, and so on. In such cases, only the arches at each end need to be anchored firmly.

† Even more ingeniously, some anonymous Roman builder found how to combine the mortar—in Latin *pulvis puteoli*—with lime, sand, and gravel to make the first concrete. The great domes of Rome, including Nero's Golden House and the Pantheon of Hadrian are built of this remarkable substance, which is more elastic than clay when wet, and stronger than granite once it dries. The concrete domes of Rome would not be surpassed in size until the age of steel.

By early 534, the piers were in place, the brick arches under construction, and the columns and cornices—huge slabs of marble, two feet by eighteen feet—which had been ordered from virtually every working marble and porphyry quarry in the empire, started to arrive. Anthemius could visit the Julian Harbor on the southern seawall of the city to accept delivery of "the white stone from the quarries in the Proconnesian islands . . . green cipollino from Carystus in Euboea, verde antico from Laconia and Thessaly, Numidian marble, glinting with the gold of yellow crocuses, red and white from Caria, white-misted rose from Phrygia, porphyry from upper Egypt."[23] One legend of the eight porphyry columns that stand in the colonnades at each of the nave's arcades has them transported from the site from the Temple of Diana at Ephesus, itself one of the Seven Wonders of the Ancient World. Another says, just as firmly, that they are from the quarry of the famed mountain of porphyry in Upper Egypt, and once stood at the Temple of the Sun near Ba'albek. Still another describes them as part of the dowry of a Roman widow who sent them to Justinian "for the salvation of my soul."[24]

As with all the elements they had specified for their structure, Anthemius and Isidore had both structural and esthetic reasons for their choices. Structurally, the alternating columns and cornices distribute the weight of the galleries and arcades to the floor, and serve as ties between the main piers; the colonnades link the primary structure, resting on those massive main piers with their buttresses, to the other "working" structure. Their esthetic value may have been even more important to the architects; the marble columns that line the arcades are of four different varieties: white from Marmara, yellow from Algeria, red from Anatolia, pink from Siga. The eight green columns from Thessaly taper slightly as they reach the ceiling, a technique—old as the pyramids—that the Greeks called *entasis*: A slight bowing outward to correct for the optical illusion that makes perfectly straight sides seem to curve inward. In their beauty and their provenance, nothing like the columns had been seen in the empire in four hundred years.

Anthemius and Isidore had another innovation in mind for their columns. Normal practice was to place columns on the second level directly above those on ground level, which was not only traditional, but by turning two columns into one long cylinder, seemingly more stable. The two architects instead placed more, but smaller diameter, columns on the colonnade level, "not with two columns, but with six Thessalian ones" wrote Paul the Silentiary, wondering "at the resolve of the man who upon two columns has bravely set six and has not shrunk from fixing their bases over empty air."[25]

The columns on the colonnade level are not merely smaller in cross-section, but in height, a "mere" fifteen to sixteen feet high, as opposed to the twenty-eight-foot monsters on ground level. Some have suggested that the architects' decision to forego stacking the columns one atop the other was economic, the result of a shortage of the money needed to pay for the larger ones. The source for the financing of the building has always been slightly uncertain; a legend from only fifty years after the church's completion describes the just-in-time discovery of a cache of coins that permitted the construction to proceed—a good story, but probably false. There is no evidence that Justinian ever questioned the cost of the rebuilt Hagia Sophia; apart from the structure itself, its ambo (the lectern from which the Gospel was read) was decorated with huge quantities of precious stones, its altar made of solid gold. One legend has the entire income from Egypt devoted to the construction, though more prudent accounting reveals that the confiscated estates of the rebellious senators beggared after the Nika rebellion was probably sufficient. So while it is certain that far less marble was being quarried in the mid-sixth century than was the case four centuries previously, which probably resulted in a relatively greater expense, the more persuasive reason for placing six relatively slender columns over two massive ones was the architects' desire to give a "greater lightness and openness to the gallery."[26] As anyone viewing the gallery can attest, in this they succeeded.

The real challenge of building a second level is that everything has to be lifted forty feet above ground level in order for work to proceed. Primitive cranes, with their masts held at the vertical by guy ropes, were the preferred means of lifting heavy loads, such as marble, but most of the bricks would have been carried up the corner ramps next to each pier by members of the ten-thousand-strong work force.

By mid-535, the time had come to put the piers and columns to the test: constructing the arches and vaults that would roof the nave. The strength and stability of an arch is largely determined by the ratio between an arch's depth—the dimension perpendicular to the direction of the arch—and its length point to point; over the same distance, an arch made of dominoes is more stable than one made of dice. The main arches, which needed to bridge forty feet from main pier to secondary pier, were built of very large bricks—more than two feet square, and nearly a foot deep, giving the arch a depth of one-twentieth of its span. Though the engineers possessed no mathematical tools for calculating stress, they did have hundreds of years of rule-of-thumb experience, which told them a depth of as little as one-one-hundredth of its span was workable. The margins were more than adequate.

Or so they thought.

Creating a beautiful and useful building that will stand for centuries is a difficult enough task. But an architect must also ensure that its unfinished structure—at each stage of its construction—is nearly as stable while being built as it will be when finished. This requirement is easy enough to fulfill when building a wall by piling stones one on top of the other.[27] An arch, however (to say nothing of a dome), is a different story; naturally unstable until complete, it must be shored up, usually by centering with a wood framework that temporarily absorbs the stress. Holes and depressions in the bricks of Hagia Sophia reveal that such frameworks were used on the main arches, but were evidently insufficient to prevent the first crisis in the building of their church. Those massive piers, intended to support hundreds of tons of brick and stone, began to give way, and they began to do so even before the arches were complete. The architects, fearful that the piers would be unable to hold, informed Justinian, who according to Procopius, told them to "carry the curve of the arch to its final completion, 'For when it rests upon itself, it will no longer need the piers beneath it.' "[28] This diagnosis is incorrect, though it is unclear whether it was the lawyer-turned-historian or the emperor himself who was displaying his unfamiliarity with the technical issues. But the problem was a real one. Because of the incredibly rapid pace of the construction—only three years had passed since the destruction of the Theodosian church; by comparison, consider that Christopher Wren spent more than five years *planning* the rebuilding of St. Paul's after the Great Fire of 1666—the stress was being transmitted to the piers before the very slow-setting mortar used to hold the bricks together on the upper parts of the piers had dried.

Justinian was correct in one respect: Nothing would be helped by deferring completion of the arches, whose condition would only worsen in their half-built state; once complete, the stresses of the arches would be transmitted evenly to the piers, and the mortar could dry at its own pace. The crumbling piers, however, were another sort of problem, and while the path the architects used to find a solution is lost to us, the results are not. Since the crisis was caused by unwanted plasticity in the pier structures, their remedy was to stiffen the two piers in some way. Not surprisingly, the two mathematicians came up with a geometric solution, changing the shape of the buttressing arch from a pure semicircle to more of a parabola, which altered the force of the arch to a more vertical direction (reducing the angular stress on the piers). Just to be safe, they also added stone projections at the gallery level and above, which strengthened the piers in the same way adding rungs stiffens a ladder.*

The redesigned arches were completed by the beginning of 536, just in time for another potential calamity: The columns supporting the north and south arches started to "throw off tiny flakes, as if they had been planed."[29] Despite their monolithic appearance, and great potential strength, marble and porphyry columns can only use the full extent of that strength if they are set into place with the same orientation as the stone had when it was originally bedded in the quarry.

---

* Unfortunately, the buttress proved insufficient, and was the source of the structural weakness that caused the dome to collapse in 558.

Otherwise, a sufficient load can put them at risk of chipping, or even cracking, and the tilt of the arches was exaggerating the effect. Again Procopius describes a Justinianic solution: Remove those columns in contact with the arches from above, replacing them once the mortar had dried. And, again, Justinian was almost right: Calculations by modern engineers reveal that the removal of only a few columns—possibly those supporting a single window—would have been sufficient to solve part of the problem. Those same calculations suggest that the crisis of 536 caused the builders to add the bronze collars that decorate the tops and bottoms of the ground level columns as well.[30]

The problems faced during construction were a direct consequence of the different rates at which pure and applied mathematics developed. By the time Justinian was ready to commission his church, the mathematics of architecture had been evolving for a millennium, ever since the massively overbuilt columns of the Parthenon—the signature structure of early antiquity—had been designed using geometry not very much more sophisticated than that taught in a modestly ambitious sixth-grade classroom.[31] Two centuries later, the simple right angles of the fourth-century B.C.E. Acropolis had evolved into the sophisticated math of solid geometry, of Euclid's spirals and Archimedes' conic sections. And by the sixth century, despite a likely deficiency of practical experience, Justinian's architects were superbly well educated in the complex curves and geometry that would be required to bring the emperor's dream to life. The *mechanikoi* were schooled in simultaneous and quadratic equations, trigonometry, solid and plane geometry, and conics.* Their mathematical tool kit could measure the area of a surface shaped like a saddle, or the volume of a container shaped like a football—and, putting those tools to practical use, could calculate the number of mosaic tiles required to cover the interior of a curved and vaulted ceiling, or the number of spectators that could be accommodated in the Hippodrome. But while they could measure and define *space* to an almost modern degree of precision—as early as

---

* Conic sections are the geometric shapes formed by a plane intersecting a cone; a plane perpendicular to the cone's circular base describes a parabola, one parallel to the base forms a circle, one at an angle to the base an ellipse, and so on.

the third century B.C.E., Eratosthenes had calculated the circumference of the earth with an error factor of less than a hundred miles—they were still neophytes at manipulating *matter*. The engineering of the sixth century was not nearly so well developed as its mathematics.

The combination of sophisticated math and simple engineering stimulated great originality, reversing ten centuries of architectural practice. During the thousand years of Greek and Roman architecture that preceded Justinian, buildings were designed using simple and precise geometric shapes, and were then modified during construction to correct for the optical distortion that a too-perfect duplicate of geometric forms would introduce.* Having trained in far more complex mathematical forms, Anthemius and Isidore were no longer slaves to geometry, but its masters, and were therefore able to distort the "pure" geometric shapes for purely emotional impact. Later observers would call the insertion of breaks in the curve of an elliptical arch, or the addition of a peak at its apex, irrational—a deviation from the simple and pure geometry of a basic conic section. In retrospect, one can see that these departures were intended not to fool the eye into believing that the form had been retained, but that it had been violated. The architects who chose this path turned their back on classical design and marched into the future, into the patterns and forms—even the style of thought—that would define the medieval age.[32] The architectural world of what would be called Christendom would not return to a truly rational architecture until the gothic cathedrals had given way to the builders of the Renaissance.

The upshot was that while Anthemius, Isidore, and their contemporaries were superb geometers, but when it came to calculating stress, force, and the properties of materials, they were scarcely better off than their predecessors of a thousand years previous. Like them, they relied almost completely on the evidence of their eyes . . . with serious consequences for the roof of Justinian's great church.

The domed roof was, in some sense, the point of the entire exercise.

---

* This phenomenon is not limited to architecture. To be pleasing to the human ear, a piano must be tempered so that the twelve semitones within an octave are uneven.

The church's walls and buttresses were, after all, merely a means to an end, which was the same for the Hagia Sophia as for the first Neolithic lean-to: to keep off the rain. Five hundred centuries of construction innovations are largely about different ways of putting a roof between people and the elements.

Posts and beams were clearly not an option. They function rather well for small buildings—houses for example—but present more and more limitations as those buildings get larger. Height is not the problem; the prime stress on a vertical post, a column, is the weight of the roof, which is the same at ten feet or a hundred feet. Moreover, while a column is compressed at its base, many materials are extremely strong in compression. As mentioned above, the horizontal beams that span a given space are a different matter altogether; they absorb greater and greater stress the longer they get. This constraint matters little even for monumentally large buildings if their monumental character is only to be appreciated by outside observers. The Great Pyramid offers a distinctly poor ratio of usable interior space to exterior surface, and even the spectacular exteriors of buildings like the Parthenon and the Temple of Karnak surround interiors that are largely forests of columns holding up stone roofs. This sort of architecture was quite satisfactory for the Egyptians, and with some exceptions, to the Greeks. It was left to Rome to perfect the dome.

The debate about the inspiration for Anthemius and Isidore's dome is now more than fifteen centuries old. In one corner are those who believe that the model was Roman, based on the theory that Anthemius visited his brother, the physician Alexander, while the latter was, according to the sixth-century historian Agathias, practicing medicine in Rome. There, goes the argument, Anthemius was exposed to structures such as the Baths of Agrippa and Nero.[33] Other historians find just as persuasive the eastern (what used to be called oriental) antecedents found in octagonal-domed buildings at Firuzabad and Sarvistan.

Both are probably missing the important point. Suggesting that the Dome of the Serapeum of Hadrian's Villa, with its alternating flat and concave panels, is a direct ancestor of the Hagia Sophia's ignores the inconvenient fact that the earlier dome simply sits flat on the drum. It lets in no light.

Anthemius and Isidore wanted to do more than simply place a roof over all that porphyry and marble; they needed to illuminate them. And to do that, they had to construct a circular dome that would sit atop a longitudinal basilica while admitting large amounts of natural light. The supports for such a dome are one of the greatest architectural innovations of all time, and one that the architects had mastered during their time in Antioch: the pendentive.

The pendentive is an inverted triangle that can simultaneously serve as a pier, as the base of an arch, and as the support for a dome of any size.

Placing a pendentive at each corner of the nave permitted the monumental size demanded by the needs of the greatest church in Christendom's greatest city, but squared off and lengthened the base over which it stood. Each of the triangular keystones hung seventy feet in the air, holding a massive dome that loomed another ninety feet higher as if it were weightless. For by the magic geometry of the pendentive, all of the weight of the roof was transmitted into four piers less than twenty feet square, leaving room for light to enter through forty windows around the rim of the dome.

In the event, though, the geometry of the pendentive proved more magical than that of the dome itself. In May of 558, as if in reproach to the architects' confidence that mastery of geometry could overcome ignorance of materials, the original dome of Hagia Sophia collapsed, its mortar unequal to its weight. And though the now-chastened Isidore rebuilt it in 563—the dome that still stands today—an eyewitness reports that it "did not strike spectators with as much amazement as before."[34]

Nonetheless, and despite the revenge that intractable bricks and mortar had on abstract mathematics, the marriage of the geometry of the dome and the physics of light was a fruitful one. By day, in both the original and rebuilt dome, the lunettes and windows poured the Anatolian sun diagonally into the nave; by night, chains suspended from the dome formed a metal circle hung with silver disks that reflected the light from thousands of lamps and candelabra. An awed contemporary observer wrote, "From twisted chains, they sweetly flash in their aerial courses . . . ships of silver, bearing a flashing freight of flame, and plying their lofty courses in the liquid air instead of the sea."[35]

Once the chains were installed, along with the altar, the ambo, and the other furniture, Anthemius and Isidore could report that their commission was fulfilled. Five years, ten months, and four days after construction began on the site formerly occupied by the churches of Constantius and Theodosius, the Church of Holy Wisdom, the Hagia Sophia, prepared to celebrate its first service of worship.* For most of the nearly six years preceding, a crew comprising one thousand foremen and ten thousand workmen had redeemed an investment estimated to be as much as 320,000 pounds of gold, but their haste to complete the final structure had resulted in careless work still visible today: Unpolished stone faces, slabs of marble simply butted together, rather than assembled with mitered edges.

On December 27, 537, Menas, the Bishop of Constantinople, departed the church of Hagia Anastasia in a chariot drawn by seven pairs of horses. As he traveled down Constantinople's broad boulevard, the emperor joined him on foot, and together they arrived at the atrium of the newly dedicated cathedral. No one present on that day recorded the details of the service, but it is certain that Justinian followed the liturgy meticulously on the day that confirmed his status as one of history's great builders. Emperor and patriarch together offered a prayer at the narthex, accompanied by the recitation of a psalm. As the two senior celebrants entered through the central doors, the congregation followed through the doors to either side; in addition to the city's aristocracy, the congregation that day would have included an imposing number of clergy, since the cathedral itself employed 60 priests, 100 deacons, 40 deaconesses, 90 subdeacons, 110 lectors, 25 psalmists, and 100 doorkeepers.[36] The cathedral's choristers alternated chanting a

---

* Compare, for example, the thirty-two years spent constructing Chartres cathedral, or the thirty-five years needed to build St. Paul's.

verse and refrain—the technical term is "antiphonal chant"—as the entourage marched down the center aisle, with the great columns to either side, and proceeded to the middle of the nave. There they faced the jewel-encrusted ambo, and behind a screen, the sanctuary with its altar of solid gold. Once there, Menas placed an enormous bound copy of the Gospel on the ambo, turned to greet the congregation and his emperor, and took his seat on the *cathedra,* once a simple bishop's chair, now a patriarch's throne made of twenty tons of solid silver. Justinian and thousands of his subjects listened to readings from the Old Testament, from an Epistle of St. Paul, and finally from the Gospel Menas had brought in to lay on the great lectern.

To that point, the service at the great church was open to all. But the celebration of the Eucharist was a mystery of such enormous significance that witnessing its presentation was reserved only to baptized members of the Christian communion. Before the body and blood of Christ could be consumed by his church, Menas formally dismissed those not yet able to receive them, and closed the doors of the church behind them. As the bishop marched past the great porphyry columns for the third time that day, the words of the twenty-fourth Psalm echoed through the vast space of Hagia Sophia's nave: "Lift up your heads, O ye gates; and be lifted up, ye everlasting doors; and the King of Glory shall come in." The congregants exchanged the Kiss of Peace, and took communion. Less than a half-mile from the besieged palace where, six years earlier, the emperor had been preparing to flee his capital, Justinian knelt, and received the sacraments.

Combining enormous intelligence and energy, driven to establish himself as the rightful successor to Constantine and Augustus, and presented with the opportunity to give tangible form to his dream of reuniting east and west, "he also raised at Constantinople many sacred buildings of elaborate beauty, in honour of God and the saints"[37] including such architectural masterpieces as the Church of St. Sergius and Bacchus. He restored the Senate House and the Baths of Zeuxippus, destroyed in the riots, and rebuilt the Great Palace; the new passage to the Hippodrome, which Belisarius had been barred from entering on the last day of the insurrection, was newly decorated with mosaics illustrating Justinian's conquests. Likewise, Anthemius and Isidore were to become among the age's most honored architects: "Anthemius designed wonderful works both in the city and in many other places which would suffice to win him everlasting glory in the memory of men."[38]

But nothing Justinian went on to commission, nothing Isidore designed or Anthemius built, truly compares with Hagia Sophia. Only a very few buildings in all of architectural history—Christopher Wren's St. Paul's, Brunelleschi's cupola in Florence, St. Peter's, and Chartres; not to forget either the Great Mosque of Damascus or the Dome of the Rock in Jerusalem, which are Hagia Sophia's more direct architectural heirs—can be mentioned in the same breath, each of them the signature achievement of an entire age. Seeking comparisons on the day of his triumph, Justinian looked to the past with his famous, if apocryphal, comment: "Solomon, I have outdone thee."*

---

* Supposedly, Anicia Juliana had included in St. Polyeuctos an inscription that suggested she believed she had surpassed Solomon.

CHAPTER FIVE

"Live Honorably, Harm Nobody, and Give Everyone His Due"

533–537

There should be one code of law for all mankind,
and one political organization.[1]

—Tatian

I N RETROSPECT, THE emperor's invocation of Solomon at his moment of triumph seems almost excessively apt; Justinian, like the biblical monarch, was to leave an even larger historical imprint with his judicial wisdom than with his buildings. Indeed, while the Hagia Sophia remained a church for nine centuries, the Justinianic Code is still taught as the founding document of legal theory to thousands of aspiring lawyers every day in the European nations destined to emerge from the shadow of Rome's empire. And, just as Justinian shares the glory of his church with his chosen architects, so, too, does he share the fame of his Code with another.

Tribonian was one of the leading peers of Justinian's empire, with titles and honors as great as any, but his stature looms even larger 1,500 years after his death. A cameo of Tribonian, sculpted by Brenda Putnam, hangs on the north wall of the chamber of the U.S. House of Representatives, next to one of Justinian himself, and surrounded by twenty of the greatest lawgivers in human history, including Hammurabi, Solon, Napoleon, and Thomas Jefferson. In 532, when Tribonian became one of the targets of the Nika rioters, he was already one of the emperor's chief counselors, and had been on the path to eminence—although some, and not just the Blues and Greens who called for his ouster, would say infamy—for most of his forty years.

As is almost always the case with even eminent personages of late antiquity, Tribonian's birth and early life are poorly documented. From his writings, we can infer that Tribonian was probably a native of Pamphylia in Anatolia; from the fact that he was junior in rank to John the Cappadocian, which suggests he was younger as well, the best guess is that he was born between 485 and 500. He was educated at one of the four great law schools of the Eastern Mediterranean, either Alexandria, Caesarea, Constantinople or, most probably, Berytus, modern-day Beirut, "the most famous of all law schools,"[2] and before arriving in Constantinople, practiced before the court of the Praetorian Prefect of the East. In 528, he was selected by Justinian to serve on the first commission tasked with creating a new Imperial Constitution, which was completed in 529. Not coincidentally, under this new legislation two senior members of the commission were removed, and Tribonian succeeded to the Quaestorship.

By the fifth century, the empire's legal and judicial functions were divided among three separate bodies, with the office of the Quaestor responsible for legislation, the Masters of Offices (there were three, supervising letters, records, and petitions) responsible for legal administration; and the Praetorian Prefect responsible for execution of the laws. Thus, Tribonian was, by the time of Nika, the chief legislative official for the entire empire. Since, in the words of the Goth historian and courtier Cassiodorus, "The quaestorship necessarily involves close familiarity with the sovereign's ideas, so that the holder can correctly express what he knows the latter feels. He sets aside his own views and clothes himself in the sovereign's, so that his words seem to proceed from the latter,"[3] Tribonian was also effectively Justinian's alter ego.

Like all of Justinian's lieutenants, he was by training and temperament a product of both Greek education and Christian faith—a son of Athens and Jerusalem. Four centuries after his death, Christian orthodoxy was so uncomfortable with its pre-Christian philosophical forebears that a tenth-century lexicon known as *The Suda* could describe him as a "Hellene"—for which read "pagan"—and atheist, a foe of every aspect of the Christian faith. Accusations of paganism would dog the reputations of many of Justinian's contemporaries, including Procopius, but in Tribonian's case they are wildly mistaken. In 531, Tribonian authored a regulation that required that before any trial or hearing could begin, everyone, including litigants and officials, was obliged to swear an oath of Christian faith while placing a hand on a copy of the Gospels . . . a requirement made easier by another regulation that ordered a copy of the Gospels placed in every courtroom.[4] Moreover, though he was a great and eloquent admirer of his pagan forebears, it is highly unlikely that an atheist Tribonian could have survived Justinian's many attacks on paganism.

Procopius described the Quaestor as "in educational attainments inferior to none of his contemporaries; but he was extraordinarily fond of the pursuit of money and always ready to sell justice for gain. Therefore every day, as a rule, he was repealing some laws and proposing others, selling off to those who requested it. . . ."[5] A belief that Tribonian's decisions could be purchased was not the exclusive property of the historian; it is, in fact, the only reasonable explanation for the demands of the Nika mob. In less than two years service as Quaestor, Tribonian had issued a new ruling on the law of inheritance roughly every two weeks,[6] and it was widely believed that the reason for the Quaestor's erratic output was not prolixity, but corruption. Even had he been completely scrupulous, however, the sheer number of decisions would have made it impossible for the possessors of any large fortune to keep up with the laws. And even had his decisions been both scrupulous and infrequent, the decision to exempt some estates from taxes would inevitably have seemed arbitrary to those not exempted. Tribonian had thus made enemies of dozens of the city's wealthiest and most influential citizens, and it seems likely that the crowd crying for his blood (or at least his dismissal) on that January day shared Procopius's judgment, or were being paid by those who did.

Justinian's dismissal of Tribonian at the height of the riot could easily have been the best-remembered fact about the onetime official, who would in consequence have become one of the thousand footnotes to the history of late antiquity. To understand why he is not it is necessary to also understand the unique importance of the law in Roman history.

———

By the time of Justinian, Rome had been home to a body of specialists in the interpretation and procedures of the law for nearly a thousand years, dating back to the time of the semilegendary Twelve Tables. The Tables were written in Rome (on ten, later twelve, bronze tablets) in 451 B.C.E. by ten of the leading city's citizens upon their return from Athens, where they had studied the work of the sixth-century B.C.E. Athenian reformer and legislator Solon.

From such romantic beginnings were born the first lawyers, as Rome, unlike Greece, treated the interpretation of law as a profession. The law that they interpreted comprised both statutes written by legislators and precedents decided by judges, but even in republican Rome, the Senate was not much of a law-writing body, issuing fewer than thirty statutes in nearly four hundred years. As a result, the law in practice was largely formed out of decisions that settled disagreements between citizens, who were obliged to bring their disputes before an annually elected official, the *Praetor Urbanus.* * He was charged with editing their stories into a form that would permit the disputants to refer their problem to a mutually agreed fellow citizen, rather like a modern judge charging a jury. The technical name for the charge was a *dicta,* and the rules that each *praetor* applied that referred to those of his predecessors became a body of "edictal law." Since *praetor*s could either ignore or forget those precedents, the first jurists—most famously the orator Cicero—rapidly became, during the last days of the Republic, indispensable. Dozens of great jurists appeared during the early imperial period, but four—Julian, Papinian, Gaius, and Ulpian—stand out in all subsequent writings. Of the four, the most quotable is Ulpian, an epigrammatist who bequeathed to his successors such nuggets as *"Ius est ars boni et aequi"* ("The legal art is properly concerned with the good and the equitable"), *"Veram nisi fallor philosophiam non simulatam affectantes"* ("The lawyer is neither

---

* Foreigners living under Roman rule brought their cases to the *Praetor Peregrinus.*

philosopher nor orator, but one who offers true opinions"), and—one of Justinian's own favorite citations—*"Juris praecepta sunt haec: honeste vivere, alterum non laedere, suum cuique tribuere"* ("The commandments of the law are these: live honorably, harm nobody, and give everyone his due").⁷

The works of the four great jurists, and those of dozens of lesser legal decisions, formed the precedents from which current disputes were, in theory, to be settled. In practice, however, their frequent disagreements on matters such as rules of evidence, or allocation of responsibility for civil wrongs meant that some rules were needed to determine which jurist's notion would prevail. Some of those rules were, to put it kindly, arbitrary; a fifth-century Law of Citations declared that:

1. Whenever the four agreed unanimously, their agreement would have the force of law;
2. When they disagreed, the opinion of the majority would prevail;
3. And when no majority existed, Papinian's ruling would prevail.

In the words of A. H. M. Jones, this legal algorithm is "justly regarded as the low-water mark of Roman jurisprudence."⁸

After a few centuries of having been cobbled together with such rules, the structure of Roman law had taken on the look of a house to which every generation of occupants added rooms without ever looking at the place from the outside; hallways that didn't connect with one another, stairways going nowhere. The confusion engendered by two emperors legislating only in the part of the empire in which they reigned simply accelerated the process. Even without an overarching dream of reunion with the west, the empire urgently needed a new code.

By the time of Justinian's accession, the empire had witnessed a number of attempts to collate the thoughts of the great jurists, the laws of the Republic, and all the enactments of the various emperors since Augustus. They include the *Codex Gregorianus*, promulgated around the year 300, the *Codex Hermogenianus* (365), and the *Codex* of Theodosius (438). In addition, Rome's "barbarian" allies produced several different legal codes of their own, of which the most famous is the *Lex Romana Visigothorum*—the Roman Law of the Visigoths, created by Alaric II in 506 just before he and his people were ejected from France. Alaric's production of the *Lex Romana Visigothorum*, which granted his subjects the same legal rights as Theodosius had granted his own in 438, was very likely the tipping point of influence on Justinian's ambitions for legal reform. Bad enough that a barbarian army had supplanted a Roman one in Roman territory, but far worse that they might supplant Roman law as well. In 528, accordingly, Justinian sponsored his First Law Commission, headed by John the Cappadocian and including Tribonian.

In Justinian's own words: "We have decided to grant now to the world, with the help of Almighty God, and to cut short the prolixity of lawsuits by pruning the multitude of enactments contained in the three Codes of Gregorius, Hermogenes and Theodosius, as well as those promulgated after the publication of his Code by Theodosius of divine memory . . . by compiling a single Code which shall bear Our own name . . ."[9] He wrote truly, for while the first attempt at a new codex was an important work, it was concerned with (mostly) the minutiae of lawsuits, with guidance in litigation. The Second Law Commission was established in December of 530 with the stated objective of revising the way in which lawyers were educated. But with Tribonian leading the commission, a "different spirit"[10] took over, that of a scholar concerned with the resolution of legal questions left unanswered by previous jurists.

In three years, that spirit produced two books: the *Digest* (a legal textbook in the form of an anthology of legal writing from the days of the Republic through the fourth century) and the *Institutes*, originally an updating of the work of the same name written by Gaius, the imperial jurist with the greatest influence upon Tribonian.[11] The magnitude of the achievement, the core of what would be known as the *Codex Iuris Civilis**—the Code of Civil Law—or the *Codex Justinianus* looms large along three different dimensions.

---

* The entire Code includes, in addition to the *Digest* and the *Institutes*, the *Novels*, a collection of laws and amendments produced by Justinian himself after the completion of the first two books.

The first is the sheer scale of the project, which eventually required a giant share of the empire's top legal minds: not merely the Quaestor (Tribonian himself) but the Master of Petitions and four of the eight men in the entire eastern empire who then held the title of professor of law. In addition, eleven lawyers were drafted from the staff of the Praetorian Prefect for the project, which was estimated would take ten years, and so represented a huge allocation of scarce resources. And, at least at the beginning, these resources probably seemed barely sufficient; the text of the *Digest* alone runs to more than eight hundred thousand words, contained in fifty books, which meant that meeting the dates demanded by Tribonian's schedule obliged each committee to turn in a draft of two books each month. To do so, the commission would have to read and examine something like two thousand other "books" comprising more than ten million words.

And they did. The text that established the objectives of the Second Law Commission asked for "a great and eternal memorial" (*in maximum et aeternam rei memoriam*) invoking the aid of emperor, minister, and God. At first glance, this reads like the sort of inflated but empty rhetoric that traditionally accompanies the benediction offered by clergy at the opening of modern legislatures or christening of modern ships. It was anything but. Contemporaneous accounts saw the hand of God in the successful completion of the task in only three years, and the achievement still seems providential. The *Digest* was completed at the end of March 533, and the *Institutes* on November 21 the same year. Both became effective on the last day of December, almost three years to the day from when the Second Law Commission was chartered. Consider that the creation of the King James Bible, which is, like the *Codex*, that rarest of literary achievements, a masterpiece composed by a committee, took seven years and is a tiny fraction of its size.

Evaluating the intrinsic quality of the work, the second dimension along which the Code casts its enormous shadow, is somewhat more difficult. The greatest achievement of Tribonian and the Second Law Commission was in remodeling the syntactical jigsaw puzzle that Roman law had become in the hands of previous generations of jurists.

Every discipline inevitably develops its own private language, and the law has always done so with the enthusiasm of lodge members exchanging secret handshakes. By the late empire, however, lawyers were using language not to exclude outsiders, but to defeat other initiates. Over the centuries, mastery of Roman law had come to be synonymous with mastery of the complications of the Latin language—classical Latin regularly used six different cases in the declension of nouns and adjectives: ablative, accusative, dative, genitive, nominative, and vocative. As a result, success at the bar of justice was frequently determined by which advocate was more skilled at matching a noun to a verb. Simplifying the practice of law so that its syntax and grammar were driven by logic rather than arbitrary rhetorical conventions was clearly an advance.

Despite this, subsequent readers frequently found the language of the *Digest* and *Institutes* to be highly grandiloquent, even pompous. Tribonian's great biographer, Tony Honore, argues that this is not a fault, but a deliberate attempt to turn the new law into a quasi-liturgical language, one that would make the subjects "conscious of the greatness of the age in which they were living."[12] Once again, a comparison with the writing of the King James Bible—for that matter, all preceding versions of the Bible—is apt. The hieratic style of the Hebrew Bible, which, in English, results in all the "begats" and "verilys" that were already becoming antique at the time of the King James translators, is a conscious attempt to reproduce a sacred language that, in its original Hebrew, was never intended to be read in the vernacular. Grand aims demand grand style, whether the aim is eschatological or judicial, and A. P. Entreves is not alone in his belief that "it is no exaggeration to say that, next to the Bible, no book has left a deeper mark on the history of mankind."[13]

And pompous or not, it is impossible to read the language of Tribonian and his colleagues without being impressed with the frequency with which the compositions that form the *Codex* refer to liberty as an ideal. Or the regularity with which equality (among Christians, anyway) is invoked as a way to eliminate dozens of forms of discrimination—between manumitted slaves and freemen, for example, or between men and women. Before the publication of the *Insitutes*, Roman women were severely restricted in owning property, were prohibited from inheriting, and were subject to lifetime guardianship. Either because of the imposing example of Theodora, or out of a core belief in equality, the *Codex* proved to be one of the largest leaps ever in the liberation of women.

More important—to Justinian and Tribonian—than removing the shackles figuratively worn by the empire's women was reattaching them to the empire's churchmen. The *Institutes* and *Novels* regulate the activities of bishops, abbots, monks, and priests, penalizing not simply behavioral offenses like gambling or theater going, but doctrinal heresies as well. In this, the *Codex* was simply the latest assertion of emperor over episcopate. To sweeten the pill he was forcing the bishops to swallow, however, Justinian agreed to add language to the Code that made Church property inalienable; that is, the lands of the Church could never be sold, or bequeathed. As a result, it was virtually impossible for the Church to dissipate its wealth in the manner of even the wealthiest families. The *Institutes* codified what would later become known as the legal doctrine of *mortmain*: the "dead-hand" that never loosened its grip on its property. Even more important, the Church retained the services of the freemen who, once having farmed the same piece of land for thirty years were forever bound to that land, thus securing not only permanent ownership of land, but of hands to work it.

On balance, therefore, the Church did well out of the work of the supposedly pagan-friendly Tribonian. The loss of Church autonomy would be recouped in less than a century, while Church prosperity would survive for a millennium; it is literally impossible to imagine feudal Europe without the great churchly estates. The bishops did not get everything they asked for; the Code continued to permit divorce—the Church's attempt to ban it was rejected on the grounds that it would increase the potential for poisonings.[14] But it was violently opposed to heresies and apostasies, and even more hostile to practices associated with pagans who had never accepted Christianity, most particularly homosexuality, which carried the most onerous of punishments, including public torture and mutilation prior to mandatory execution.

Neither the language of the *Codex* nor its religiosity proves most troubling for "modern" lawyers . . . that is, those living after the Renaissance. The most lasting criticism of Tribonian is not bad writing or severity, but fraud.

Tribonian's charge was emphatically not the creation of an entirely new legal code. On the contrary, in order to command the respect of the citizenry at large and the legal profession specifically, the new *Codex* needed to trade on the already existing deference granted to the earliest jurists. Thus, a typical entry from the *Digest* reads:

1. Theft is a dishonest handling of a thing in order to gain by it, or by its use or possession. Such conduct is against the very law of nature.
2. [Gaius] There are two degrees of theft: manifest and non-manifest.
3. [Ulpian] A manifest thief is one whom the Greeks describe as "caught in the very act."[15]

The style is explicit: This is what Gaius said, this Papinian, and so on. But, in dozens if not hundreds of cases, what Tribonian and his colleagues incorporated is *not* what Gaius wrote, but what they wanted him to have written, in order to be consistent with Ulpian, or Julian. Had the 1954 U.S. Supreme Court followed the style of Tribonian, *Brown vs. Board of Education* would not merely have overruled the "separate-but-equal" endorsement given by its predecessors forty-eight years before in *Plessy vs. Ferguson*, but altered the words offered in *Plessy* itself.

The commission was trapped between the rock of precedent and the hard place of consistency. The application of precedent—a fundamental value to anyone who respects the law—meant they had to incorporate existing statutes and prior decisions; but the contradictions within and between those statues and decisions still needed to be eliminated. The modern solution is to add the original decisions to legal registers with the words "as amended," and modern critics of Tribonian take him to task for his failure to do the same. However, this misunderstands the sixth-century perspective. Tribonian did not regard judicial opinions as in any way different from statutes . . . and a current statute book does not incorporate all the laws that have ever been in force, even for historical purposes. Once it was added to the *Codex,* even a private legal opinion assumed the status of legislation, and like the legislation was subject to subsequent amendment.

However, the likelier argument is found in the remit of the Second Commission: not to collect, for further study, all previous laws and interpretations, but to filter out all the detritus and admit only currently valid law. The commissioners were forced to either cite anonymous precedents, or to put words in the mouths of predecessors that were not previously there. Thus, perversely, the respect that the commission gave to historical precedent trumped their respect—if, indeed, they had such—for scholarly honesty. Unlike previous compilations, like that of Theodosius, the work of Justinian and Tribonian found a way out of the trap that made earlier efforts merely a pasting together of old laws; the *Digest* needed not to compile, but to legitimate, the laws. When they did not agree, they were *made* to agree.

Despite Justinian's mania for attaching his chosen name to geographical and political enterprises without number—at least twenty-seven separate cities were named for him,[16] including his birthplace in Macedonia, which was renamed Justiniana Prima and made a bishopric—only one survived the century after his death: the *Codex Justinianus.* The real reason for the status still accorded Tribonian and Justinian in legal history fifteen centuries after completing work on their Code is neither its length nor its coherence. It is the influence that it continues to exert on the law—the third, and most important, of the criteria by which the importance of the Code should be judged.

The influence reaches substantially further than the minutiae of legal proceedings. It is impossible to comprehend either the glories or the ignominies of the political entities that would become modern Europe without understanding the great theoretical structure created by Tribonian and his colleagues and known today as the Civil Law tradition. The Inquisition and the Renaissance, the Napoleonic Code and the Holocaust are all, in their ways, an outgrowth of the *lex regia,* the idea that found its final and fullest expression in Tribonian's masterwork: *Quod principi placuit, legis habet vigorem*: "The will of the prince has the force of law."[17]

Given this weight of significance, it seems improbably perverse that the *Codex* would almost certainly have died in infancy but for its fortuitous rediscovery five hundred years after it was written—one of the sixth century's longer-gestating children, to be sure, but one that still appeared in time to shape the medieval world. Sometime during the eleventh century, a huge portion of the *Codex*, including the only surviving manuscript of the *Digest*, which was owned by a monastery in the Italian city of Pisa since the 700s, was itself copied and brought to the great medieval university just becoming established in the city of Bologna. There, the first notable legal scholar of the Middle Ages, Imerius, established a school whose members were collectively known as Glossators. These scholars were rabid in advocating for a more coherent replacement for the codes that had, by default, become the legal structure of what had been the western half of Rome's empire: Alaric's *Lex Romana Visigothorum* and the chaotic assortment of religious-themed statutes that had begun to be enforced by the papacy under the name Canon Law. The Glossators, the first to rename the Code the *Corpus Iuris Civilis*, found a ready audience for it among secular rulers, which in retrospect seems scarcely surprising. By establishing the position of the emperor—by extension, any anointed ruler—as the legislator of divine will, the *Lex Sacra*, the Code was an essential, perhaps *the* essential, endorsement for what would eventually become the divine right of kings.[18] Justinian's regnal successors may have respected the Code's breadth, scope, and internal consistency, but they *loved* the legitimacy it gave to their kingly prerogatives.

The Civil Law, as it came to be known, remains central to European legal systems to this day. Though the preeminence it now gives is not to kings but to legislatures, the basic principle is identical: In any dispute at law, the institution that drafted the statute prevails. The Civil Law tradition, however, despite its enormous significance, remains best known in the English-speaking world as a comparison—

usually an invidious comparison—to the other great legal tradition, that is, the Common Law.

In the Anglo-American system known as the Common Law, the institution that *interprets and adjudicates* the statute has the final word. In its most extreme form, America's highest court asserts a right of judicial review over even the laws produced by the national legislature, but everywhere that the Common Law prevails—essentially Great Britain and her former colonies—case law frequently trumps legislation. Partly this is a function of the very different way that the Common Law evolved: incrementally, selecting rules and procedures one at a time, rather than building everything according to a grand design. One practical consequence is the very different rhythm of a Common Law trial, in which parties to a dispute enter written pleas, one after another, until the various charges and countercharges can be distilled into a single material issue, after which questions of law are decided by a judge, questions of fact by a jury.[19] The same parties, in a civil law proceeding, offer an entire case to the presiding judicial authority, who then must rule on both matters of law and matters of fact.

This would seem a minor distinction, of interest only to legal scholars and professionals, except for one thing: Because of the deference granted to whomever writes the laws, and the absence of a jury to rule on matters of fact, the Civil Law tradition is fundamentally friendlier to tyrannical regimes than the Common Law. The nineteenth-century U.S. Supreme Court Justice Joseph Story believed that the judicial works of continental Europe "abound with theoretical distinctions, which serve little purpose other than to provoke idle discussions and metaphysical subtleties, which perplex, if they do not confound, the inquirer. . . ."[20] As Lord Coke put it, under the Common Law, every man's house is his castle; not because it is defended by moats or walls, but because while the rain may enter, the king cannot; under the Civil Law, the king is bound by nothing at all. Justinian left the European nations that grew out of Roman soil more than just the law; he bequeathed them autocracy, as well.

Even so, the founders of the American nation, though heirs to the Common Law, were quick to honor the document that is central to the Civil Law tradition. When a commission led by James Madison made a list of recommended acquisitions for use by the first Continental Congress, the first two items were Justinian's *Institutes* and the *Corpus Iuris Civilis* (Blackstone's *Commentaries* appeared in fifth place). One should take care not to see the two traditions as inimical; what they share, which is an acknowledgment that rule by law requires adherence to *some* preexisting body of mutually agreed precepts; the obligation to follow precedent in the common law (in legal Latin *stare decisis*) has its analogue in obedience to the statutory authority.

Moreover, the philosophical line dividing the Civil and Common Law, though real, is not as sharp as some texts suggest. In September of 2005, when a group of U.S. senators were evaluating the fitness of Judge John Roberts to be the seventeenth Chief Justice of the United States, their questions turned on precisely the same issues that confounded Tribonian: What is the appropriate level of deference given to settled legal precedent? To the prerogatives of the legislature? And—most important—who is to decide? In the English Common Law system, judges maintained a consistent hold on their prerogatives for hundreds of years, and were therefore able to assert authority over time, just as, for Justinian, the imperial will was not supposed to be recreated anew with each change of emperor.

———

After he was discharged under duress by Justinian in order to appease the Nika rioters, Tribonian's titles were removed. Despite this, his work continued, work that "constitutes one of the most brilliant feats of organization in the history of civil administration." Whether anyone else could have replaced Tribonian in 530 C.E. we cannot know, though the constitutions composed by preceding and succeeding Quaestors give no evidence of a comparable intellectual drive; in the words of Tony Honore, "the future of European legal culture hung perhaps by a single human thread."[21] In the event, the thread of Tribonian's life proved strong enough, and Justinian's loyalty robust enough, that the author of the Code was to spend only a modest time in disgrace. Only two years after his discharge, Tribonian was formally rehabilitated, and once again appointed to the Quaestorship. He resumed his office, and his position as Justinian's lawyer, and maintained both until his death. The date of that death is unknown, but clues from Procopius suggest that Tribonian had passed away by mid-544.

By then, the legal position of the emperor, under Justinian, had achieved the same apotheosis as the law itself: in the *Codex,* for the first time in history, the emperor is named *nomos empsychos* which translates as "law incarnate," which is more than Augustus or any of his successors ever achieved. Bending the law to his imperial will, however, did not increase his respect for it; the opposite, in fact. In the prefatory note to one of the laws written during the latter years of Justinian's reign, the emperor who would be remembered more for his legal code than for any of his other achievements, wrote "Laws do for the business of life what medicine does for diseases; consequently the effect is often the opposite of that intended."[22]

CHAPTER SIX

"The Victories Granted Us by Heaven"

533–540

"When I treat with my enemies," replied Belisarius, "I am more accustomed to give than to receive counsel; but I hold in one hand inevitable ruin, in the other peace and freedom."

—Gibbon, *The Decline and Fall of the Roman Empire*

O F ALL THE pieces of the legal code that bear his name, the ones in which Justinian's own hand—as distinguished from that of Tribonian, or any of the great historical jurists—are most evident are the *Novels*, the new laws and interpretations issued during the emperor's own reign. In 539, more than five years after the defeat of the Vandal Kingdom, Justinian "wrote," "It was out of enthusiasm for this [emancipation] that we undertook such extensive wars in Africa . . . *for the freedom of our subjects*"[1] (emphasis added). Whether the law offered a rationale or merely a rationalization for the reconquest of North Africa, no one can doubt that Justinian believed fervently in the rightness of his cause.

The residents of Libya had been Roman citizens for seven centuries, but had been embarrassingly free of imperial authority since the first decades of the fifth century. In yet another object lesson in the law of unintended consequences, Vandal swords had cut the links that bound Libya—like Spain, another spoil of the Punic Wars—to the empire. Those consequences date from 427, when the Visigoths hired by the empire to return Spain to imperial control had done the job so well that the only Vandals remaining were confined in Andalusia (the name, originally Vandalusia, is one of the bequests they left behind). In that same year, Boniface, the Roman military commander in Africa (the more accurate term is "generalissimo"—the fifth century was dominated by such men, whether "barbarian" kings or nominally Roman officers),[2] was recalled by the emperor, his loyalty suspect. He refused, prompting the empire, probably at the instigation of Boniface's rival, the general Aetius, to send two punitive expeditions against the rebel. In self-protection, Boniface arranged for an alliance with the Vandals, offering a Libyan kingdom in payment. As a result, in 429, eighty thousand Vandals[3] under Gaiseric, the most formidable of the Vandal leaders, "excellently trained in warfare and the cleverest of men,"[4] crossed from Spain into Africa. Either unaware of her general's complicity in the Vandal invasion, or willing to overlook it, Galla Placidia pardoned Boniface and ordered him to defend the Roman province that he had deliberately put at risk.

His defense was somewhat inglorious. Following a successful siege of the city of Hippo in eastern Algeria (during which the city's bishop and most famous resident, St. Augustine, died) the Vandals conquered Carthage in 439, and attacked Sicily shortly thereafter, defeating Boniface several times along the way. In light of this, Procopius's tribute to Boniface—that he "attained to such a degree of high-mindedness and excellence in every respect that if one should call [him] 'the last of the Romans' he would not err, so true was it that all the excellent qualities of the Romans were summed up in [him]"[5]—seems at least questionable. When pardoned, he turned on his newfound friends, discovered himself besieged, tried to fight his way out, was "badly beaten"[6] and, humbled, returned to Italy.

Boniface's rival, Aetius, purchased a larger share of future historical attention by leading the combined Roman and Visigothic army to victory over Attila's Huns at the Battle of the Catalaunian Plains, near Chalons, in 451. As a result of blunting the Hun invasion at its farthest extent, Aetius became for a time the most powerful man in the entire western half of the empire. That time ended quickly, however; a year after the death of Attila, in 454, the emperor executed his most formidable general.

With Boniface in disgrace and Aetius dead, Rome offered little resistance to the militarily emboldened Vandals, who had acquired skills at sea to match their confidence in battle. Their maritime expertise made them nothing less than an imperial nightmare; a law of 419 promised death to anyone teaching "sea matters" to barbarians.[7] Rome's concern was well founded. By 453, Gaiseric was raiding the Italian peninsula, and in 455 he entered Rome at the head of a looting expedition, returning to North Africa with plunder both very old—the gold-and-bronze roof from the temple of Jupiter, dating back to Domitian and Titus, for example—and very young: the princess Eudocia, daughter of Valentinian III.

Over the course of the next fifteen years, Gaiseric's Vandals—by now his confederation included Alans and Suevi as well; in fact, as Procopius tells us, "the names of the Alani and all the other barbarians, except the Moors, were united in the name of Vandals"[8]—continued to discomfit both halves of the empire.

The east, however, remained robust enough to do something about it. In 470, the emperor Leo assembled, in Constantinople, an enormous army to lead a punitive expedition against the Vandals, with the ultimate object of restoring imperial authority in North Africa. Unfortunately, the army was better supplied than led; Leo's general, Basiliscus, made a landfall off the North African coastline less than twenty miles from the Vandal base in Carthage, but despite an overwhelming numerical advantage, delayed his landing by nearly a week, either from an excess of caution or a deficiency of integrity—Procopius reports that Gaiseric bought the delay by bribing the general. In the event, the delay was well purchased, since it lasted long enough for the wind to change in a way favorable to the Vandals' experienced navy. Gaiseric's admirals towed unmanned fireboats into the Roman fleet, destroying it, the army it carried, and Leo's dreams of conquest. Freedom from imperial rule would survive even the death of Gaiseric, seven years later.

Another of Gaiseric's legacies would survive him as well—this one more unfortunate for his people. The Vandals' law of inheritance required that each king's oldest direct male successor automatically succeeded to the throne upon the death of his predecessor. In the year

523, the plan showed its underlying weakness, as Eudocia's son Hilderic, despite the fact that he was prouder of the ancestry he could trace to orthodox Roman aristocrats than to Arian Vandal royalty, became the new Vandal king. Apparently, he had few if any other qualifications for the job, and, in 530, he was deposed, in favor of his more enthusiastic cousin, Gelimer.

Hilderic may have been lacking in allies among the Vandal leaders, but he did have a friend in Constantinople. Even before Justinian had succeeded Justin as emperor, he had been corresponding with Hilderic—they "made large presents of money to each other"[9] according to Procopius—and the deposed king knew exactly where to turn for support against the usurper, Gelimer. Though it took some years of posturing on both sides, during which the new emperor warned Gelimer to forego exchanging "the title of king for that of tyrant,"[10] he agreed to send an army in support of Hilderic. The expedition was far smaller than the one sent by Leo: fifteen thousand regular troops* plus one thousand "barbarian" allies. But what it lacked in size, it more than made up for in leadership. In command was Belisarius.

———

In the third week of June, 533, Justinian's armada, numbering 500 troop transports, plus 92 escort ships, manned by 32,000 sailors,[11] departed Constantinople's harbor. The ten thousand infantry and five thousand cavalry included veterans of the Battle of Dara such as Aigan and his Huns, and Pharas and his Herulians, and a number of other lieutenants, including Dorotheus and Solomon, the latter a eunuch. Perhaps most surprisingly, the army also included the general's wife, Antonina.

Had she lived in any time other than that dominated by the figure of Theodora, Antonina would surely qualify as the most fascinating and formidable woman of her day. Daughter of a theatrical prostitute and a professional charioteer, "she reigned with long and absolute power over the mind of her illustrious husband; and if Antonina disdained the merit of conjugal fidelity, she expressed a manly friendship

---

* That is, from the *comitatensis,* or as it was starting to be called, the *stratiotai.*

to Belisarius, whom she accompanied with undaunted resolution in all the hardships and dangers of a military life."[12] At the beginning of the Vandal War, at least, the resolution of Antonina was much on display. The general's wife was, like her best friend, Theodora, a woman of voracious sexual appetite and formidable intelligence, but it was the latter rather than the former that figures in her first appearance in the historical record. Before even leaving the Aegean, the fleet was delayed by an episode of food poisoning, which Procopius blamed on the war-profiteering of John the Cappadocian. The historian goes so far as to accuse the Prefect, Justinian's prime minister, of adding water to inflate the weight of the promised "twice-baked" bread known as hardtack, causing it to rot. And if the expedition had too much water in its bread, it had too little in its water kegs, which were leaking so badly that they were essentially useless. While little could be done about the bread, however, Antonina deserves credit for helping the fleet to escape the consequences of a water shortage; her idea—storing reserve water in glass bottles buried in sand—may not have saved the expedition, but it came close. But even with its immediate logistical problems solved, the Africa-bound force was still heavily outnumbered and about to confront a martial nation on its home ground.

Belisarius was acutely aware of the tactical and strategic challenge he had undertaken, and knew that victory required the cooperation of the Libyans, who, after all, had been Roman citizens for centuries longer than they had been subjects of the Vandals. Stories too many to list testify to Belisarius's honor—and his charisma; no Roman general since Mark Antony cut a more dashing figure—but the most prominent characteristic of this remarkable soldier may have been his pragmatic intelligence. He had a highly developed sense of justice, but an even keener eye for military advantage. In light of this, the general's decision, while en route across the Mediterranean, to charge, convict, and publicly impale two Huns for public drunkenness looks less an opportunity to assert discipline for its own sake than as an object lesson to his other troops, particularly on the matter of looting and rape. To the general, two soldiers were a small price indeed if it bought the loyalty of the Libyan citizenry.

Belisarius rarely forgot any detail that might have been important to a mission. As the fleet made its way to Libya, the general sent Procopius on an intelligence-gathering side trip to Syracuse, in Sicily, where the historian-turned-secret-agent met with one of his own countrymen, suborned him, and discovered that no naval ambush was waiting for the Roman arrival. Free of worry, Belisarius and his 15,000 men landed on the coast of North Africa at Caput Vada, five days march from Carthage. In August of 533, the successor to Scipio, the Roman general who defeated Hannibal, had returned to Africa.

And he was there with the same goal in mind: Carthage, the Vandal capital, toward which Belisarius and his army marched in full battle order and gear, with cavalry scouts two miles in front of the main body of troops, the Huns under Aigan two miles to the army's left, and the Mediterranean on his right. His caution seemed prescient when, at Decimum—so named because it marks the milestone ten miles from Carthage—the scouts discovered a Vandal army in hiding, and turned a potential ambush into a Roman victory. But the ease with which Belisarius's small expeditionary force swept north to Carthage is only partially explained by good planning. More than eighty years previously, the Vandal king Gaiseric had torn down the walls of every city in Libya save Carthage itself, in order to prevent their ever serving as either a rebel base or an enemy's advance garrison. Gelimer's ancestor had thus inadvertently made it possible for an aggressive general to maintain offensive pressure without ever being bogged down in a siege, which proved to be an important advantage for Belisarius, who traveled without even the rudiments of a siege train and needed to maintain mobility at all costs. His sixth-century version of the *blitzkrieg* rolled into Carthage only four weeks after landing at Caput Veda . . . so quickly in fact, that Gelimer was barely able to evacuate his capital before Belisarius entered it, accompanied by Procopius, who wrote "and it happened that the lunch made for Gelimer on the preceding day was in readiness; and we feasted on that very food and the domestics of Gelimer served it and poured the wine and waited on us in every way."[13]

The loss of Carthage left Gelimer without a capital, but not without resources. Now on the outside looking in—and looking in at new fortifications being built every day by the Roman conquerors—he planned an insurgent campaign against the occupiers. In need of allies, Gelimer sent a frantic appeal to Sardinia, an island controlled by the Vandals since 455, where several thousand of his best troops, under his brother Tzazon, were trying to suppress a rebellion. When Gelimer and Tzazon learned that the revolt had been fomented by Justinian through his agent, the governor of Sardinia,[14] the brothers attempted to return the favor by promoting rebellion among the Arians who made up a significant portion of Belisarius's federated troops, and particularly the never-very-well domesticated Huns.

By December, Tzazon and his troops had arrived from Sardinia and joined with Gelimer, assembling an army twenty miles from Carthage, at the village of Tricameron. The size of the combined Vandal army is variously estimated as between 50,000 and 150,000 men—between three and ten times the size of Belisarius's force, which was further weakened by the refusal of the Huns to fight, a result of Gelimer's successful attempts to suborn them—but despite this numerical advantage, it was unable to cope with the disciplined Roman cavalry. Three successive charges by Belisarius's mounted bowmen against Gelimer's numerous but unblooded troops, were decisive the day before the Roman infantry (led by that remarkable woman, Antonina) even arrived. Tzazon was killed, the Vandal forces broke and ran, and, to all intents and purposes, North Africa was once again a Roman province. Justinian memorialized the event in the introduction to the *Institutes* of 533: "Africa and other provinces bear witness to our power having been, after so long an interval restored to the dominion of Rome and to our Empire by the victories granted us by heaven."[15]

Even so, Belisarius's victory seemed less a validation of Roman military might than an indictment of his Vandal opponents, whose ferocity and skill had declined greatly from the days in which they terrorized Rome itself. The army that had slaughtered Leo's expedition had, in the intervening seventy years, not merely failed to adopt the tactics that Belisarius's cavalry had learned fighting the Huns and Persians; they had also gone soft. From the moment that the Romans landed in North Africa, Procopius encountered and recorded examples of what he must have regarded as the corrupting effects of civilization, beginning with the historian's surprised admiration for the Vandals' beautifully cultivated fruit trees at Grasse—today Sidi Khalifa, and still famous for its gardens. The Vandals may have been barbarians, but they did not live barbarically; a century of African living had produced no military challenges, but it did accustom them to daily baths. The Vandal kingdom was no longer a match for Belisarius's professionals and were essentially defeated in less than a month of campaigning.

The one threat that still required resolution was Gelimer himself. After Tricameron, the last Vandal king escaped to Numidia, where he established himself in a mountain aerie of the Phoenician-speaking tribes Procopius called Moors, whom he believed to be descendants of the "Gergesites and Jebusites," refugees from the conquest of biblical Canaan by Joshua.[16] There, just above Hippo, the onetime bishopric of St. Augustine, he was besieged for nearly three months by Herulians under the command of Pharas, the hero of Belisarius's victory at Dara. The Vandal king's Moorish hosts were not as accustomed to civilization as his own people, and Gelimer, a man of some musical and literary pretension, actually spent part of the siege composing an ode bemoaning his lack of a sponge. When he finally surrendered, he evidently did so believing that it was preferable to be a clean slave of Justinian than an unwashed king of the Vandals.

The glory of Belisarius's triumph was in no way diminished by the ease with which it was gained, and upon his general's return, Justinian determined on a public appreciation. The most appealing choice for an emperor supremely conscious of his historical connection to Augustus and Constantine was a triumph, the ceremonial parade of plunder and noble prisoners that preceded the passage of a victorious general in review before his sovereign. In Rome, triumphs were so rigorously managed that, for centuries since Augustus honored Agrippa after his campaign in Spain, they had always followed the same route through the city: from the Circus Flaminius, through the Porta Triumphalis, across the Circus Maximus, around the Palatine Hill and up to the Capitol.[17] In contrast, Belisarius, in 534, walked directly from his home to the Hippodrome, preceded by the much traveled spoils of Gaiseric's 455 sack of Rome. Occupying a place of honor was a solid gold, seven-branched candelabrum that the emperor Titus had brought to Rome after his first-century campaign against the fractious occupants of the imperial province of Judea: the Jews.

––––––––

Many aspects of the Mediterranean world in the time of Justinian are dependably exotic to modern eyes, but one of its more familiar characteristics is the ubiquity of the Jews. By some accounts no fewer than 10 percent of the citizens of the sixth-century empire may have been Jewish, but even more striking than the raw numbers, to the modern reader used to medieval and modern images of Jews-as-types (the merchant, the scholar, the physician), was the sheer ordinariness of encountering Jews in virtually every walk of life. Late antiquity is a time when Jewish fishermen, actors, farmers, and sailors appear regularly in contemporaneous accounts. The fifth-century bishop of Ptolemais, Synesius of Cyrene, describes a comic, though awful, voyage from Alexandria to Palestine aboard a small vessel with a Jewish crew . . . whose captain, also Jewish, abandoned the tiller at sundown on Friday, despite a storm that nearly destroyed the ship.[18]

It was not always thus. For millennia, the Jews were far more closely tied to their national cultural home than the Romans were to Rome, and the breaking of those ties, whether by forced emigration or voluntary assimilation, proved even more traumatic for the Jews than the eastward shift of the imperial capital was for the empire. The journey of the menorah Belisarius carried to Constantinople in 534— which had once stood guard over the temple built in Jerusalem* by exiles returning from Babylon a thousand years before—is a proxy for the Jews' entry into the transnational life of the Mediterranean. While in exile in Babylon, after their defeat and displacement by Nebuchadnezzar, they had not built temples; Psalm 137—"by the rivers of Baby-

––––––––

* Built, as it happens, on the original site of the Temple of Solomon, the apocryphal inspiration for Justinian's proud boast on the completion of the Hagia Sophia.

lon we sat down and wept"—describes their inability even to sing the melodies central to Jewish worship when barred from Jewish land. So after the Persians defeated the Babylonians and permitted the return of the exiled Jewish aristocracy, including the members of the hereditary temple priesthood, the first order of business was a rebuilt and refurnished temple, including the candelabrum that liturgy demanded be lit and cleaned only by priests. Through war, rebellion, and even a brief conquest at the hands of the Seleucid Greeks, the temple and the menorah were the religious focus of an entire people, and it remained so when that people started to spread outside Judea. Even the first-century philosopher Philo of Alexandria, a thoroughly Hellenized Jew, was obliged to contribute one-half shekel annually to the high priests responsible for upkeep of the great temple. Those high priests, the *kohanim*, were responsible for more than temple upkeep; for as long as Judaism remained a hierarchical state religion, they were their people's absolute monarchs. And so they remained, until the first-century revolt that ended with destruction of the temple, defeat at Masada, and confiscation of the spoils of Jerusalem.

The victorious general and future Emperor Titus (whose Triumphal Arch in Rome features an image of the stolen menorah, and through which Jews have long refused to pass) had destroyed the temple in the hope "that the Jewish religion might be utterly eliminated," but he was to be denied—and not merely denied, but rebuked—as Rome's empire proved, over the subsequent centuries, essential to Jewish survival. Despite Rome's desire to obliterate the Jews' national identity (to the point of renaming Jerusalem as Aelia Capitolina, and Judea as Palestine[19]), "the very agent of Jewish defeat in Palestine was also the agent of Jewish revitalization as an autonomous Diaspora, a commonwealth within an empire."[20] Roman arms may have destroyed Jerusalem's temple, but Roman law enjoined toleration of other religions. The empire offered protected status to Jews and the houses of worship they had built during their Diaspora: the synagogues.

During the centuries from Titus to Justinian, the synagogues, like the menorah, traveled across the Mediterranean. The travels changed them. Temple Judaism depended upon the sacramental activities—

ritual garments, blood sacrifices—of a class of hereditary priests, the only ones granted privileged access to the Holy of Holies inside Solomon's Temple. The farther that the religious nationality built around the temple traveled, in time and space, from Jerusalem, the more it evolved into one dominated by professional scribes who came to dominate Jewish life. By the third century, the scribes had transcribed the body of oral law known as the Mishnah; by the end of the fifth century, rabbis working in Jerusalem had produced the first of the two books of commentary called the Talmud; by the end of the sixth, in Persia, the other.

As the textual mastery exhibited by the scribes grew in value, the importance of heredity declined. From the year 85, Rome had administered the Jews of the Diaspora through the patriarchate, a family dynasty that retained some of the leadership functions of the high priests of the old temple, including the collection of taxes, which the patriarch then remitted to the empire. But during the 430s, as the Vandals began their conquest of North Africa, and Attila succeeded to the leadership of the Huns, the position of the patriarch went unfilled, thus turning local governance of the Jews—of 10 percent of the population of the Roman Empire—over to the scribes, or as they had by now become known, the rabbis.

So when the seven-branched symbol of the Jews' still powerful connection to their temple preceded Belisarius on his march into the Hippodrome in 534, Justinian's Jewish subjects recognized it for what it was. In 1936, the Austrian novelist and playwright Stefan Zweig imagined a scene in which the rabbis of Constantinople prevailed upon the emperor to return the menorah to its home in Jerusalem:

> "Lord, your rule, your city are at stake. Be not presumptuous, nor try to keep what no one yet has been able to keep. Babylon was great, and Rome, and Carthage; but the temples have fallen which hid the [Menorah] and the walls have crashed which enclosed it. . . ."
>
> After thinking matters over for a few moments, Justinian said dryly: "So be it. Let the thing be taken from among the spoils of Carthage and sent to Jerusalem."[21]

The emperor had a well-deserved reputation for religious intolerance—although Jews did not suffer from it in any way like they would in medieval Europe, they were nevertheless subject to occupational and professional restrictions; construction of new synagogues had been banned since the time of Theodosius; and rabbis were forbidden to teach within them, by Justinian's Novel 146[22]—but he evidently did plan to allow the recovered menorah to be placed in a Jerusalem church, as evidence of the triumph of Christianity. The menorah, confiscated by a pagan general, stolen by Arian barbarians, was dispatched, by a Christian emperor, on the final leg of its journey home.

After his conquest of Gelimer and acquisition of the menorah, Belisarius had claimed another spoil of the Vandal campaign in the name of his emperor: a Sicilian fortress, nominally ruled by the Ostrogoths. The response of the Ostrogoth ruler—"friends settle their differences by arbitration, enemies by battle. We therefore commit this matter to the Emperor Justinian to arbitrate in whatever manner seems to him lawful and just."[23]—is a model of temporizing diplomacy. Its author was yet another of the remarkable women whose exercise of political power echoed through the sixth century.

Her rise to such power is an historical accident. Though Justinian was as skilled in diplomacy as Belisarius was in warfare, even he was frequently stymied by the unexpected. The emperor had engineered the adoption by Justin of Theodoric's son-in-law, Eutharic, even naming the Goth-king-in-waiting a Roman Consul as the first step in a peaceful rapprochement with the Ostrogoths. His plan was knocked off kilter when Eutharic died unexpectedly in 522, and his widow, Amalasontha, niece of a Frankish king and daughter of Theodoric, became ruler of virtually all of Italy as regent for her son Athalaric.

Amalasontha was just as much a Romanophile as her husband had been, though far less able to resist the demands of the Goth nobility. The Ostrogothic "barons" were determined that their future king be educated not in Roman, but Gothic style, which evidently included considerable training in brawling and carousing. Amalasontha was forced to give up her son, but she refused to give up her regency, and began the correspondence with Justinian to which she referred in her response to Belisarius's claim on the Sicilian fortress.

That correspondence took on more urgency once the queen-regent felt herself in danger from the intrigues forming at the Gothic court at Ravenna. In response, she initiated a conspiracy with the emperor: In return for the emperor's protection, Amalasontha proposed bringing her entire treasury—some three million gold *solidi*, more than ten tons of the precious metal—across the Adriatic to the Illyrian city of Dyrrachium (now in Albania).[24] The emperor agreed, but in early 534, before Amalasontha could make her escape, her son Athalaric died of what Gibbon called "premature intemperance." Her claim to the throne gone, the resourceful woman proposed an alliance with her cousin Theodahad—an alliance he accepted, marrying the Ostrogoth's queen-regent later that year.

Amalasontha had accurately perceived the danger of the Gothic court, but misidentified its source: Theodahad had married Amalasontha only to betray and imprison her in a fortress on the island of Martana in Lake Bolsena in the mountains of Tuscany. Martana became the scene for the last act of Amalasontha's story, but its particulars remain murky. That the onetime queen of the Ostrogoths was murdered in her prison is unquestioned; not so her murderer. Gibbon's description of Amalasontha—"At the age of about twenty-eight years, the endowments of her mind and person had attained their perfect maturity. Her beauty, which, in the apprehension of Theodora herself, might have disputed the conquest of an emperor, was animated by manly sense, activity, and resolution"[25]—clearly derives from the accusation of Procopius and others that Amalasontha was killed at the behest of Theodora. The empress was reportedly fearful of both her allure and growing closeness to Justinian, and sent Theodahad a message that no one would object if his cousin were executed.*

---

* Some plot lines have proved so common in historical narratives that they seem to suggest the existence of inexorable laws, or at least narrative conventions, governing human history. One favorite is the story of the beautiful and sophisticated princess foiled by a conspiracy of her own barons and a jealous neighboring queen. The well-remembered story of Mary Stuart, the French-speaking, Catholic queen of Scotland betrayed by Presbyterian lairds and judicially executed by Elizabeth I is eerily anticipated by the life of Amalasontha.

For a historian attuned to the politics of jealousy (and in this Procopius has few equals), the year of Amalasontha's death was a rich one. That same year, Belisarius's wife, Antonina, developed a passionate infatuation with her adopted son, Theodosius, which she apparently consummated at every opportunity once she returned with Belisarius from Africa. So careless was she that her chambermaid, Macedonia, informed Belisarius of his wife's infidelities. The results, as reported by Procopius, were predictable: After the general learned of the affair, he sent soldiers to kill Theodosius. After the soldiers lost their quarry, Belisarius flew into a fury against Antonina. After his fury spent itself— the implication is that he was diverted by sex; the general's single weak point seems to be his wife's undoubted allure—he forgave her. After he forgave her, he revealed the name of his informant, and after Antonina learned the name of her betrayer, she cut out Macedonia's tongue, chopped the rest of her into small pieces, and threw them into the Mediterranean. However, as with so many of Procopius's splenetic stories about Theodora and Antonina, which he wrote but never published (in a volume entitled the *Anekdota*, or *Secret History*, only discovered in the seventeenth century), a word of caution is in order. Procopius was the most conservative of men, by birth a member of the petit nobility, by temperament a Hellenist, by profession a lawyer, and, in consequence, a man who valued historical precedent above everything. The mere presence of such clever and resourceful women so close to the levers of military and political power was undoubtedly troubling; that they were women who were born to families of circus performers must have been odious. So even if Procopius's instincts as a historian prevented him from *inventing* libels against Antonina and Theodora, the temptation to *publish* those that were already in wide circulation would have been very real, indeed. In the event, the *Secret History* reveals at least as much about its author, and his world, as it does the targets of his vitriol.

––––––––

Whether Theodora had reason to be jealous of Amalasontha may, in the end, be as irrelevant as the truth about Antonina's fidelity. Her death gave Justinian the pretext to invade Italy, nominally a territory

whose king owed allegiance to the emperor, but practically an independent Gothic kingdom since Theodoric. The emperor was determined to convert the fiction of imperial authority into reality, and his
tools were to be the two soldiers who had preserved that authority
during the Nika Revolt. In the fall of 535, Justinian ordered Mundus,
the *magister militum,* or master of soldiers in Illyricum, to advance
into Goth-held territory in Dalmatia, and sent Belisarius to Sicily at
the head of an even smaller expeditionary force than the one he had
led to victory against the Vandals.

Before he could reach Sicily, however, the general was diverted
south. The troops in Carthage commanded by Solomon, the lieutenant whom Belisarius had left in charge while enjoying his triumph
in Constantinople, had mutinied. Solomon had escaped, along with
Procopius, and brought the news to Belisarius while he was en route
to Sicily, and the general immediately headed for Africa. There, he
was able to restore order, and with one hundred chosen men of his
comitatus, returned to his by-now restive army aboard ship on the Sicilian coast.

By December, Belisarius's four thousand soldiers had surprised and
taken Palermo, in another exhibition of his remarkable mastery of
both the geometry and the psychology of the battlefield. Realizing
that the fortress (the strongest on the island still held by the Goths)
was inaccessible from the landward side, he sailed his fleet into the
harbor with a corps of archers occupying the highest masts of each
ship, a novel method of applying the classic infantry objective—to
take the high ground—to naval tactics. In possession of the advantage
that gravity gives to missile troops positioned higher than their opponents, Belisarius predicted that the garrison would panic, which they
did, surrendering the fortress. With Palermo under control, he was
able to take Syracuse, as well, essentially returning Sicily to imperial
control.

Even before the fall of Palermo and Syracuse, the presence of an
imperial army in Sicily, however small, had motivated Theodahad to
reopen peace negotiations with Justinian, which was very likely the

strategic rationale for the invasion. In late 535, however, a Gothic army in Dalmatia killed Mundus in battle, and Theodahad's spine stiffened. So did Justinian's, who ordered a new army into Illyricum, and Belisarius into Italy. In early spring of 536, therefore, Belisarius led his troops across the Straits of Messina into the boot of Italy, heading up the west coast of the peninsula, keeping his fleet on his left. By summer, he was 120 miles from Rome, at the gates of Neapolis, the "New City" built in the shadow of Mount Vesuvius: Naples.

Naples was to prove a test of both Belisarius's resolution and his cleverness. For weeks, the city's residents fought doggedly against the invading Romans, none more ferociously than the Neapolitan Jews, whose bellicosity was fueled less by loyalty to Theodahad's Goths than by fear of Justinian's persecution. Of this, they had much experience, since the administration that Belisarius had left behind in North Africa was, like the general's emperor, determined to impose orthodoxy wherever it was threatened. In the former Vandal territories, not only had all antiorthodox Christians been banned, but the region's synagogues were rapidly being turned into churches.

High walls and determined resistance combined to make Naples impregnable, until one of Belisarius's private soldiers discovered that one of the aqueducts that had formerly supplied the city's water was no longer in use, and as a result was unguarded and accessible outside the city's defensive perimeter. Belisarius ordered a team of sappers to enlarge the aqueduct's diameter to the point that it could accommodate a single file of armed and armored soldiers; this they did, under the very noses of the city's defenders, keeping noise to a minimum by using scrapers rather than picks and shovels, and opening a route under the city's walls.

Because the aqueduct was covered, however, no one knew where it would finally permit entry into the city. After weeks of tunneling, several hundred of Belisarius's finest soldiers crawled nearly a mile into the very center of Naples, with no idea where they were until they reached a point where the aqueduct was open to the sky. The access point was twenty feet above ground, next to a private home occupied by a single woman, who was taken captive by the troop while they rigged harnesses to climb down from the elevated aqueduct, took the north tower of the city, and ended the siege. The response in Constantinople was enthusiastic; Justinian wrote "We have good hopes that God will grant us to restore our authority over the remaining countries which the ancient Romans possessed to the limits of both oceans and lost by subsequent neglect."[26]

The reaction in Gothic Italy was somewhat less happy. After the fall of Naples, the Ostrogoths finally grew tired of the inconstant and unmartial Theodahad; within a month he was deposed, and by December, killed by one of his own nobles and replaced by another, named Vitigis.

December 536 was a busy month for Italy: On the ninth day of the month, Belisarius, responding to an invitation in which Silverius, the bishop of Rome (only retrospectively was he to be referred to as the pope), agreed to open the city's gates to him, marched north along the Latin Way and entered the Eternal City through the Asinarian Gate, at the southwest corner of the city. It signifies little that the army was mostly made up of Illyrians, Huns, and Moors, or that its leader was a Thracian whose allegiance was to an empire whose administrative center was closer to the Black Sea than the Mediterranean. Nor does it matter that Rome's population was only a fraction of its size during the age of Augustus. Sixty years after the overthrow of Romulus Augustulus, Rome was again part of the empire.

———

How long it would stay an imperial city was problematic. Belisarius's three thousand mounted horsemen—two thousand of the original force had been detailed for garrison duty in other Italian cities—faced no opposition when they entered Rome, whose Ostrogothic garrison had already departed.* But holding it was another story. The city's defensive perimeter, largely defined by the walls built by Aurelian during

* Or mostly departed; the Goth's commander, one Leuderis, remained behind to formally surrender the keys to the city . . . literally. He, and the keys, were sent by Belisarius to Constantinople.

the third century, was twelve miles in extent, which meant that even if every soldier was manning a post, he would be defending more than twenty feet of rather porous wall. Even with the moat and protectively angled walls that Belisarius immediately started building, defending the city in the traditional manner was clearly impossible.

Certainly the Ostrogoths thought so, and regarded the loss of Rome as merely a temporary setback. Vitigis almost immediately began correcting matters. His first act was to send reinforcements to Dalmatia, where imperial troops were pressing on to the coast despite the death of their commander, Mundus. He followed by summoning a force of fifty thousand Ostrogoths to his standard, and marching from Ravenna to Rome. Learning of their imminent arrival, Belisarius ordered the troops he had previously sent into Tuscany under his two chief lieutenants, Bessas and Constantinus, to return and aid in the city's defenses. Rome's first siege was about to begin.

The account of any siege is inevitably measured against the *Iliad*, and few of Homer's successors can have been more conscious of his literary ancestry than Procopius, who observed each act of the battle for the city from Belisarius's side. He produced a Homeric saga, and the care he took in preparing his narrative is due in part to the enormous symbolic importance that the city of Peter and Augustus retained even after losing much of its strategic value. To this most conservative historian, in fact, the most impressive characteristics of Rome were not the symbols of the empire, the church, or even the republic, but the great artifacts of Greco-Roman history, especially the spoils of the Trojan War, such as the Palladium, which in legend Odysseus had taken from the city of Troy, and the ship—"marvelous . . . transcending all description"—in which Aeneas supposedly fled the burning city.[27]

The home of these relics was too large to be completely invested, even by a fifty-thousand strong Goth army; however, it was *far* too large for Belisarius to completely defend with only a few thousand men. Even concentrating his troops at the city's fourteen large (and several small) gates spread them too thin, and Belisarius ordered most of them sealed. He had just completed doing so when Vitigis's army arrived, divided itself into six camps forming a rough semicircle around the northern arc of the city, and prepared for the siege's first battle.[28]

That clash took place at the same place as Constantine's most famous battle, the Milvian Bridge. Belisarius had built a tower to defend the bridge and manned it with some of the mercenaries that formed a portion of his defending army . . . not the most dependable portion, since they deserted at the first sign of the Gothic army. As a result, several hundred Roman cavalry, led by Belisarius, were surprised on their way back from a patrol and had to retreat to the walls of the city, only to find them barred. With his command and his life under threat, Belisarius, riding a war-horse with a distinctive white star on its face that, once known to his enemies, prompted dozens of Goths to concentrate fire on him, "made a display of valor such, I imagine, as has never been shown by any man in the world to this day . . . there fell among the Goths no fewer than a thousand, . . . but by some chance Belisarius was neither wounded nor hit by a missile on that day."[29]

Over the course of the next weeks and months, Vitigis and Belisarius entered into the middle game of the chess match to which all sieges are inevitably compared. Unable to mount a riverine assault after Belisarius strung a chain across the Tiber, Vitigis destroyed the aqueducts whose water powered Rome's flour mills. The attempt to starve the garrison failed when Belisarius mounted the mills on boats in the Tiber, and "fastened ropes from the two banks of the river and stretched them as tight as he could, and then attached to them two boats side by side and two feet apart, where the flow of the water comes down from . . . with the greatest force . . . so by the force of the flowing water all the wheels, one after the other, were made to revolve independently, and thus they worked the mills with which they were connected and ground sufficient flour for the city."[30] When the Goths tried to foul the mills by throwing debris into the Tiber, Belisarius extended the chain across the Tiber, catching all the potential problems upstream.

After failing to deny bread to Belisarius's soldiers, Vitigis turned toward a frontal assault on the city, directing his engineers to construct four huge ox-drawn siege towers: fortified frameworks designed to hold battering rams suspended from interior chains.* As the Goths rolled the rams toward the city, Belisarius climbed to the firing step along the facing rampart and, at great range, and to the morale-building delight of his troops, killed two Gothic officers with his own bow. The general's demonstration was less a boast than a demonstration of how to target the oxen pulling the towers, and once the imperial archers saw the point, they were able to foil the attack. When Vitigis attacked the newly fortified Tomb of Hadrian—now the Castel Sant'Angelo—Belisarius's soldiers destroyed the statuary on the tomb's roof and used them as missiles to blunt the Gothic assault.

And so it went; each time Gothic infantry attempted to scale Rome's walls, Belisarius's horsemen sortied from a flanking gate, driving off the Goth cavalry with mounted archery and then running down the helpless infantry with their lances. Even so, the overwhelming asymmetry of forces favored Vitigis to such an extent that Belisarius sent a letter to his emperor, requesting reinforcements, "for one ought not to trust everything to fortune, since fortune, on its part, is not given to following the same course forever."[31]

The modern reader should parse this carefully. When the letter, almost certainly drafted by Procopius, invokes "fortune," it means something different than simply "luck." This is actually yet another invocation of Fortuna, the Neoplatonist personage from Boethius's *Consolations* and, as such, a reminder of just how close the sixth century sits to the fulcrum of the Western intellectual tradition. It is a moment when orthodox theologians could debate the most arcane points of Christology while invoking oracular signs, Sybilline prophecies, Delphic oracles . . . and Dame Fortune.

---

* The siege of Rome, in fact, is a kind of catalog of sixth-century military machinery; in addition to Vitigis's rams, the campaign featured scorpions (Procopius calls them "ballista"; giant bows, carrying oversize arrows fletched not with feathers but thin pieces of wood), onagers (catapults built with a strong piece of wood holding a sling at one end, and the other plunged into a twisted hawser of rope; the rope's torsion imparted velocity to the arm of the engine), and "lupi" or wolves (essentially a twenty-foot-high framework of crossbeams with a sharp goad protruding from each point where perpendicular beams intersect that could be dropped like a deadly drawbridge on troops approaching a breached wall).

The attention given by Procopius and Belisarius to omens did not, however, distract them from the decidedly more mundane details of the siege. When Vitigis captured Rome's port, Ostia, he was able to finally impose a blockade on deliveries of food to Rome. As hunger set in, and promised reinforcements failed to arrive, the citizens of Rome became restive, so when Belisarius intercepted a letter that his onetime ally Bishop Silverius sent to Vitigis, in which he once again offered to open the gates of the city, the general took immediate action. The bishop—who had already made an enemy of Theodora because of his hostility to the empress's Monophysitism—was arrested and brought to Belisarius's headquarters in the Pincian Palace, there to be greeted by Antonina, who confronted him with the fact of his attempted betrayal to the Goths. The bishop was then relieved of his finery, dressed and barbered as a monk, and marched past his underlings as an object lesson in the price of treachery.

After the collapse of Silverius's fifth column, Vitigis agreed to a brief truce so that he could regroup. Belisarius took advantage of the truce to bring in supplies and several thousand reinforcements of Isaurian infantry and Thracian cavalry. When combat resumed, it had taken on a more measured scale; a few dozen soldiers on either side, mostly on horse (and therefore able to refuse battle whenever it suited them) might come into contact, but most of the encounters turned into single combat. This suited Procopius's self-conscious adherence to the Homeric style well—he devotes pages of his account to the story of one of Belisarius's guards, the Hun Chorsamantis, wounded by Gothic treachery, whose bloody revenge ended in his own death and that of a dozen or more enemies under the eyes of his comrades.

While Belisarius was perfectly capable of daring acts of single combat, as during his fighting retreat from the Milvian Bridge at the beginning of the siege, the general was always a soldier first, and a warrior second, more Odysseus than Achilles. Belisarius's soldierly virtues—his pragmatism, energy, and leadership—had kept Rome defended and fed long enough that the Goths started to suffer from famine themselves: in Procopius's words, "while they were in name carrying on a siege, they were in fact besieged by their opponents."[32]

So, having failed to take the city by storm, blockade, and treachery, and unable to supply his still large army, Vitigis now turned to negotiation.

Procopius's description of the negotiations can seem slightly bizarre, given the blood that had been shed over the preceding months. Vitigis first asserted the legal validity of the Gothic position, arguing that Theodoric was acting at the behest of Zeno himself when he established the Goths in Italy. As a corollary to this assertion, Vitigis cites the fact that the Goths passed no laws of their own, obeying all the laws of Rome in preference. Pointing out that, despite their own Arianism, the Goths had also acted in a respectful manner toward orthodox Christians, Vitigis offered Sicily to Belisarius. Since imperial troops were already ocupying the island, Belisarius responded by offering the Goths Britain, which had been independent of imperial rule for a century, with the clear implication that they might as well offer him Constantinople, or for that matter, the moon.

Though the negotiations, as a result, resolved nothing, they did convince Belisarius that the initiative had shifted in his favor. Seizing the moment, the general sent one of his lieutenants on a raiding mission to Tuscany at the head of two thousand horsemen. Though under orders to avoid large troop formations or fortifications, the officer, a soldier with the unlikely name of John the Sanguinary—"a daring and efficient man in the highest degree, unflinching before danger and in his daily life showing at all times a certain austerity and ability to endure hardship unsurpassed by any barbarian or common soldier"[33]—immediately headed for Ravenna, the Gothic capital whose garrison had been drained to support the siege of Rome.

Vitigis, learning that his capital was under attack, attempted several final assaults against Rome. The last of them, like the first, took place on the site of Constantine's battle against Maxentius: the Milvian Bridge, in a battle "equal to any that preceded it."[34] All the attempts failed, and on March 12, 538, more than a year after beginning to besiege a city defended by fewer than four thousand soldiers, the remainder of what was once a fifty thousand strong Gothic army, beset by hunger and disease, withdrew.[35]

Vitigis had left more than his capital unguarded while trying to dislodge Belisarius. His queen, Matasuntha, Amalasontha's daughter and herself a granddaughter of Theodoric, had been in Ravenna when John and his cavalry arrived, and was taken enough by the Thracian to propose an alliance "including marriage and betrayal of the city."[36] Before the two could consummate their plans, Vitigis himself arrived, to find not only John's raiders, but another imperial army, newly arrived in Italy, under the command of an Armenian eunuch who was to become the only real rival to Belisarius among the imperial generals Justinian employed in his campaign of reconquest.

Little is known of Narses's early years, or even how he became a eunuch, whether from castration as a young slave, to deliberate "preparation" by his parents, to even being born sterile.[37] There is more agreement about the physique of the Persarmenian courtier-turned-general; Agathias's "slight, frail-looking man"[38] and Gibbon's "feeble and diminutive body"[39] were regularly and disparagingly compared to Belisarius. In fact, the contrasts between the two generals—one young, tall, handsome, and passionately devoted to his wife, the other a short, unprepossessing eunuch—are almost excessively appealing to later observers, so much so that later fictionalized descriptions of Narses include everything from a lisp to a hunchback. What the two shared, however, seems more significant: a high order of intelligence, loyalty to Justinian, and a gift for war.

Narses's career prior to his arrival in Italy was far more typical of an imperial bureaucrat's than a general's. He was already in his fifties, and a power at court, when Justinian selected him as *agent provocateur* to carry bribes and disinformation to the Nika rioters, a clear indication of the emperor's trust. The trust was due as much to Narses's status as to his proven loyalty; like celibate priests, eunuchs offered their superiors a devotion uncompromised by family ambitions. Moreover, according to Procopius, Narses was also a Monophysite, and therefore probably favored by Theodora as well. The Armenian ultimately rose through the ranks of the palace chamberlains, the *cubicularii,* to the highest such position in Constantinople's hierarchy, Grand Chamberlain of the Court, from which he so enjoyed the favor of the emperor that he was admitted to the Patrician Order, and membership in the Imperial Council of State.⁴⁰

He was, however, an unlikely general, one who came to military command for the first time in late middle age. Narses was probably close to sixty when he arrived in Italy at the head of ten thousand soldiers, and while nominally under Belisarius's command, almost immediately found himself in conflict with the Thracian. The proximate cause of the clash was John the Sanguinary, still pursuing his separate alliance with Matasuntha even as her husband had him besieged in the city of Ariminum (as Rimini was then known). Though—mostly— successful in his raiding, John had also earned a reputation for "unreasoning daring and a desire to gain great sums of money"⁴¹ among many of his comrades, and had put his head in a noose in service of that desire. While Belisarius was deciding whether to rescue or chastise his subordinate, Narses spoke in John's behalf, and as a result Belisarius somewhat unenthusiastically assembled a relief column and broke the siege. The greeting offered by the emaciated John was likewise lacking in enthusiasm: He publicly thanked not Belisarius for his deliverance, but Narses, pushing Justinian's greatest generals into a lifelong rivalry.

A lack of cooperation did not prevent the two armies from taking the battle to Vitigis, and one city after another fell to Justinian's soldiers. The despoliation of the countryside and cities alike resulted in a brutal famine, even cannibalism. Procopius described starvation's victims almost clinically—"their skin became very dry, so that it resembled leather more than anything else . . . they changed from a livid to a black color [and] came to resemble torches thoroughly burned."⁴² But the greatest human disaster of the Italian wars was not starvation, but massacre. The victim was the city of Milan.

Milan had originally been founded by tribes from Cisalpine Gaul—the Italian side of the Alps; the French side was known as Transalpine Gaul—but absorbed into the Roman Republic in 222 B.C.E. Mediolanum, its original name, had grown to be an imperial capital under Diocletian, a center of religious authority under its influential bishop, Ambrose, and once the seat of the Vicar of Italy. It was also no stranger to the depredations of war, having been attacked by the Hun armies of Attila in 452, and by Belisarius's lieutenant Mundilas, who had occupied the city in 539. When Vitigis besieged Milan later that same year, Narses sent an army, under John, to relieve it. John's troops traveled from Ancona northwest toward Milan, but stopped on the south side of the Po River, intimidated by the Goth army they believed awaited them on the other side. Mundilas, the commander of the besieged garrison, smuggled a letter out to Belisarius, who ordered the relieving column to proceed, at which point John climbed another rung on the ladder of insubordination by refusing, willing to proceed only if commanded by Narses. After a plea from Belisarius, Narses agreed to take charge, but the time consumed in the passage of letters turned out to be disastrous for the citizens of the besieged city, who had been reduced to eating rats.

The only hope for Milan's civilians lay with the remaining imperial troops, who weren't eating any better. The Goth army besieging the city knew this, and crafted their surrender offer accordingly: safe passage for the garrison, but no similar assurances of mercy to the civilian population. When Mundilas, the garrison commander, received the offer, he appealed to his own soldiers to refuse it, knowing that it would inevitably result in a massacre. His oratory was, however, unequal to his sense of honor; the garrison deserted, and the Goths destroyed Milan. While Procopius's tally of three hundred thousand dead is clearly an exaggeration, other accounts make it clear that the civilians were wantonly murdered, and the second largest city on the Italian peninsula effectively, though briefly, ceased to exist.

One of the consequences of the destruction of Milan was a series of violent recriminations from both Narses and Belisarius; another was Justinian's response, the recall of Narses. Free of the distractions of a split command, Belisarius proceeded to approach Ravenna, systematically taking—by storm, or by siege—each of the Goth strongholds in the Po valley as he marched north. By the end of 539, Vitigis, having failed in his attempts to find allies, and feeling the noose tightening on his capital, finally ran out of options . . . or so he believed. In fact,

Justinian, growing impatient, had sent envoys to Italy carrying the details of a proposed peace treaty with Vitigis that would grant the Gothic ruler all of Italy north of the Po River, with the emperor retaining direct authority over everything else. There can be little doubt that Vitigis would have accepted the offer, if he had been permitted to see it.

However, Belisarius saw it first, and was "moved to vexation"[43] at the prospect of anything short of total victory. Vexed enough, in fact, to disobey his emperor's command, though, as subsequent events proved, only to satisfy the emperor's desire. Knowing that Vitigis was well and truly trapped, he waited long enough for the Goth negotiators to arrive. When they came, however, they carried an unexpected offer: The Goths were prepared to tender their surrender to Belisarius, on the condition that he agree to serve as their king. Accepting, as a *ruse de guerre*, kingship over his enemies, the Thracian general marched into Ravenna cheered by its residents as their new ruler.

———

When Procopius first saw the Gothic capital of Italy, he was initially most impressed by the "very wonderful thing [that] takes place every day,"[44] the flow of the tidal estuary that turned the city into an island every morning, and connected it back to the mainland of Italy every evening . . . a phenomenon that is more than simply curious, but a testimony to the easy defensibility of the city Honorius had made his capital once Milan started to seem too vulnerable.

More striking to most visitors, then and since, are Ravenna's buildings. The great treasures of Ravenna include the tomb of Theodoric the Great and the Church of San Vitale, which houses the only remaining images of Justinian and Theodora done in mosaic. But in some ways the most significant of Ravenna's architectural riches is an eight-sided brick building located in the city's northeastern quarter, a short distance southwest of its two cathedrals. Four of the sides feature semicircular apses, crowned with tiled roofs. History does not record whether the Thracian general who was, briefly, Ravenna's king ever visited the building, but had he walked to the northwest corner of the building, and entered its interior, he would have seen it fitted out with its original decorations, most especially its baptismal font. It was, after all, a baptistery, a place where what had once been an open-air ceremony, a simple initiation ceremony into the Christian community, had become a solemn once-a-year Easter ritual.[45] The font did not long survive Ravenna's conquest, nor did any of the baptistery's decorations, save one. Seventeen centuries after its founder was excommunicated, anyone entering can lift his eyes to heaven, and see through to the heart of Arianism.

The frescoed ceiling of the Arian baptistery of Ravenna—it is called that to distinguish it from the Orthodox baptistery built fifty years before, sometime in the mid-fifth century—presents one of the rarest of art history miracles: a simple visual representation of a complex theological dispute. In the Arian baptistery, the pictorial Jesus—nude, in water up to his waist, but with his procreative equipment quite visible—faces toward the east, while the twelve apostles approach an empty throne. Jesus thus achieves divinity, obedient to Arian doctrine, only at the moment of baptism.[46]

The Arian baptistery was fated to witness the end of Arian rule in Italy; only days after his entry into Ravenna, the general repudiated his brief kingship over the Goths and sent his predecessor in chains to Constantinople. Had Vitigis possessed a sense of architectural irony, he might have remarked on the city's most prominent statue of the emperor, an equestrian monument atop a circular column clad in bronze "softer in color than pure gold, while in value it does not fall much short of an equal weight of silver,"[47] placed just north of the imperial palace above the public square outside the senate house. Significantly, the statue of Justinian, astride a horse, holding a globe in his left hand, dressed as a Homeric warrior, was placed above sculptures of "three pagan kings kneeling before Emperor Justinian and offering their cities into his hands."[48]

The coda to the conquest of Italy was written by Procopius: "It is not at all by the wisdom of men or by any other sort of excellence on their part that events are brought to fulfillment, but that there is some divine power which is ever warping their purposes and shifting them in such a way that there will be nothing to hinder that which is being brought to pass."⁴⁹ In little more than two years, Justinian had restored the core of his empire, and, despite the renewal of conflict with Persians on the Syrian border, had a reasonable expectation of another several decades on the throne, had every reason to believe that his territorial gains would stand, with his church and his code, for centuries.

———

Even as Justinian's star seemed permanently in the ascendant, his prime minister remained the most despised man in the empire. Scarcely any figure of late antiquity is as consistently reviled as John, but neither contemporaneous critics nor subsequent historians were as hostile as his two great enemies: Theodora and Belisarius, both of whom earned his animosity out of what appears to be envy of their proximity to Justinian. It is scarcely surprising, therefore, that his downfall was the work of the person who was friend to one and wife to another: Antonina.

In the spring of 541, Antonina identified the Cappadocian's weak spot: his daughter Euphemia. John doted on the girl—in the words of one historian, it was "the one amiable trait in his repulsive character,"⁵⁰ but he had evidently given her more of his love than his shrewdness. Thus, when Antonina told Euphemia that Belisarius was not only feeling unloved by his emperor, but unrewarded for his great successes in conquering both the Vandals of North Africa and the Goths of Italy, she was sowing seed on well-prepared ground. Euphemia had been listening to her father's resentments toward the imperial couple for years, and was thus more inclined to believe any hostile words than to suspect the motives of their source. Antonina exploited the girl mercilessly, telling her that the only thing preventing Belisarius from rebellion was the lack of support from civilian ministers . . . from someone like John.

But while one can understand the daughter's lack of skepticism, the only explanation for the father's credulity, once he was apprised of the conversation, can be that his ambition had completely overcome him.* For John not only agreed to a secret meeting with Antonina, he did so on her terms, at a villa in Chalcedon. And not only did he meet in her territory, but he spoke, without inhibition, of his readiness to strike for the throne. And not only did he speak treason, but he did so within the hearing of witnesses.

Antonina, of course, had been working all along with Theodora. Once the meeting was agreed to, the empress supplied the final piece of John's downfall, for it was she who sent two men whom the emperor trusted above all others: Narses and Marcellus, commander of the palace guard. Hidden behind a wall, both overheard John's treachery, and attempted to arrest him on the spot. Though John briefly escaped their swords, their words were to convict him to Justinian, who exiled his own prime minister—and, as a result, placed him out of the reach of the microorganism that had reached a level of virulence that was about to stagger the entire Mediterranean world.

* It should be noted that while Procopius makes no criticism—in the intended-for-imperial-eyes history of Justinian's reign entitled the *Wars*—of Antonina's plan for using a man's daughter to suborn him, he is most critical (in the *Anekdota*) of her swearing of false oaths to convince both Euphemia and John that her motives were pure . . . or, at least, honestly treacherous.

A SIXTH-CENTURY SHIP *that departed Alexandria bound for Constantinople might, wind and waves permitting, make its first landfall about four to six days out, off the shore of one of the islands that make up the Cyclades chain. From there, it would travel through the strait formed by a headland on the north and the island of Nissiros to the south into a bay, from which it would travel along the coast of Anatolia until it reached the Sea of Marmara and the capital. Its total time at sea would be a generally uneventful ten days, perhaps two weeks.*

Imagine one such ship, weighing anchor during the spring of 542. During its two weeks in Alexandria's harbor, it takes on a cargo of grain, along with the usual dockyard impedimenta: rats. No one thinks twice about them. Rats, even dead rats, are as familiar to sailors as sunburn. Or fleabites.

One day out of Alexandria, one sailor complains of headache, fever, pain in his legs and back. On the second day, two more grow ill, and the first victim notices a painful swelling in his groin. The afflicted sailors become confused, their speech slurred as if drunk. Their eyes grow bloodshot, and under their skin, blood starts to pool, causing blackness in the fingers and toes. By the third day, more sicken. They hallucinate; one throws himself overboard, either because he is delusional, or merely to stop the burning temperature. No one knows, for his tongue is so swollen before his suicide he cannot be understood. By the fourth day, only one sailor is left alive. He beaches his vessel on the shores of Halicarnassus, and runs from the cursed ship, screaming . . . but not very loudly, as he has started coughing up blood. He makes it as far as the nearest village before dying.

Now, multiply this voyage by a hundred, five hundred ships. By a thousand ports. Ten thousand oxcarts.

The demon was loose.

PART III

BACTERIUM

CHAPTER SEVEN

"Daughter of Chance and Number"

It is not surprising that microbes now find us so attractive. Because the carbon-hydrogen compounds of all organisms are already in an ordered state, the human body is a desirable food source for these tiny life forms. Bacteria see us as a source of autopoietic maintenance in their ancient struggle against thermodynamic equilibrium.[1]

—Lynn Margulis and Dorion Sagan

O N A DAY more than eleven centuries after the conquest of Ravenna, and in a city six hundred miles to the northwest, a man sat down at a table, placed his eye at one end of the tube of a rudimentary microscope, and saw something no human being had ever seen before. His name was Anton van Leeuwenhoek, and he made his living as a lens grinder in the Dutch city of Delft at a time when being Dutch and manufacturing lenses were never more prestigious. During the last quarter of the seventeenth century, the Netherlands were enjoying the summer of their Golden Age, and precisely machined glass was proving to be one of the great scientific tools of all time, so much so that the secret methods used by lens grinders like Leeuwenhoek and Baruch Spinoza frequently died with their creators. From London to Venice, astronomers and admirals were paying a premium for Dutch-made telescopes, but Leeuwenhoek directed his acute powers of observation not to the wide horizons of his nation's maritime domains—a contemporary echoed Virgil in describing the seventeenth-century Dutch empire as having "no other bounds than the Almighty set at Creation"[2]—but to the densely populated world of a single drop of water. "Whenever I have found anything remarkable,"

Leeuwenhoek later wrote, "I have thought it my duty to put my discovery down on paper, so that all ingenious people might be informed thereof." In 1683, an image reproducing his discoveries appeared in the *Philosophical Transactions* of the Royal Society, a drawing of living things so small they were at the limit of the size that his lenses could resolve. Leeuwenhoek called the organisms "animalcules."

One hundred years after Leeuwenhoek, the Danish scientist Otto Friedrich Muller applied the binomial naming system of Carolus Linnaeus to Leeuwenhoek's creatures, and a hundred years after *that,* the nineteenth-century Frenchman Louis Pasteur discovered the complicity of the tiny organisms in causing disease. But it was not until 1872 that the German scientist Ferdinand Julius Cohn first named the rod-shaped creatures *bacteria,* from the Greek word for staff, *baktron.* Though neither Cohn, nor Pasteur, nor Leeuwenhoek realized it, the objects of their interest were Earth's first life form, and will almost certainly be its last. In the words of the paleontologist Stephen Jay Gould, "On any possible, reasonable, or fair criterion, bacteria are, and always have been, the dominant form of life on Earth."[3]

There is something unavoidably arbitrary about choosing the beginning of the path that connects Earth's first living thing with Rome's last great emperor . . . assuming, that is, that the path itself isn't just a literary caprice. Most historians of late antiquity have tended to scant the import of the demon—of any disease—choosing more traditional narrative elements to tell the story of classical civilization's last days. Their preference is understandable. At first glance, the tools of history seem somehow ill-matched to a story whose leading character is a creature with no sense of history, or without even a sense of self-awareness.

The first glance, however, is deceiving. One of the key insights of all the great biological scientists, from Darwin forward, is that the history of even a single-celled organism is still, after all, history, in some sense unrepeatable and unpredictable, but not unconstrained. The history of life, like that of humanity, flows through a channel confined by the edicts of nature, of course; the dimensions of a bird's wing cannot violate the cube-square law, for example. But it is also narrowed by choices made. Neither a civilization nor a species can effortlessly change a survival tactic once it is mastered, any more than a drop of water can retrace its path to the point where one river turned into two. Heraclitus's famous dictum—that one cannot step into the same river twice—is as true for evolution as it is for history. The origins of nations, like that of species, are always unique.

In all texts on the life sciences, from the rudimentary to the sophisticated, the passages devoted to bacteria are rich in the story of origins, as well they should be. Paleontologists regularly push back their estimate of the moment when Earth's tree of life first broke ground, and many of their discoveries are still somewhat controversial. But while professionals will continue to argue about the details, they long ago reached consensus on the big picture: The appearance of the first recognizable bacterial life more than 3 billion years ago, the first cells containing a nucleus about 1.5 billion years later, and the first multicellular life somewhere between 570 and 700 million years ago. Another 300 million years later, the great saurians called dinosaurs appeared, and 60 million years after their extinction, a group of bipedal primates migrated out of East Africa to eventually discover language, fire, and the itch for empire. Which means that for nearly two billion years, bacteria were not only the lords of creation; they were creation itself.

When evolutionary biologists write about the next billion and a half years, though, they tend to shift their attention to bacteria's descendants: to foraminifera, diatoms, fish, flowering plants, carnivorous reptiles, and mammals, replicating, in their own way, the cartoon image of a primordial fish crawling onto a beach and changing successively into an amphibian, a sloth, a gorilla, a Neanderthal, and finally, at the far right, into a gloriously upright modern man. To be sure, modern evolutionary biologists are usually at pains to remind readers that evolution is not "progressive" and that animals are not "higher" than plants and that an amoeba is just as much a miracle of natural selection as a blue whale. Even so, any picture of the tree of life that places bacteria at the bottom—or even, more correctly, at the center— runs the risk of forgetting something rather important:

Bacteria didn't stop evolving once their descendants showed up. They haven't stopped doing so yet.

––––––––

When the demon began the last stage of its own evolution, its immediate ancestor may have been living anywhere between the River Nile and the Bay of Bengal,* but for now, it is probably more useful to adopt the creature's perspective, and to say that it lived in a somewhat more circumscribed universe: the mammalian gut. Like all bacteria for the previous three and a half billion years, it was very small—so small that it approached the lower limit of life itself. Fifty of them, stacked atop one another, would just about equal the thickness of a dollar bill. *Yersinia pseudotuberculosis*, as it would one day be called, was, by the scorecard kept by natural selection, a highly successful organism: wide ranging, gigantic in numbers, and, in general, so innocuous in its effect on its host that it could survive for decades in the same human intestine, causing little more than an occasional flulike stomachache.

It also possessed, like all its bacterial ancestors, a gift for change. The four thousand genes in the single cell of *Y. pseudotuberculosis* included dozens of sequences that permitted chunks of genetic code to move around to different spaces on the genome—and to pick up new bits of code from other bacteria, from viruses, even from bits of DNA floating in the bacterial liquisphere like loose pages of a manuscript.

These pieces of code were useful only insofar as they offered some advantage in the deadly serious business of growing and reproducing. But because it could not know whether any particular bit of code would turn out to be helpful, *Y. pseudotuberculosis* had to rely on the law of large numbers to find a selection advantage, either by the theft of existing code from elsewhere, or from the introduction of a typo into its own codebook while duplicating it. The proverbial roomful of monkeys that will eventually type out the King James Bible required far fewer attempts than the bacterium hunting for the blueprints needed to build an entirely new organism.

––––––––

* For more on the debate about the location of the original plague basin, see pages 194–96.

And yet, new bacterial species are appearing literally every hour of every day.

Considering that bacteria have continued evolving during the billion-plus years that they shared the planet with other life forms— roughly the same time they had to evolve before any other life form turned up—it seems odd that general histories of life give such short shrift to the "second act" of bacterial evolution. The underemphasis can be laid, partly, at the feet of taxonomy. From the time of Leeuwenhoek until the middle of the nineteenth century, taxonomists, no freer than anyone else from anthropocentrism, grouped all living things into either plant or animal kingdoms, and bacteria were hammered into place in one or the other. The ones that possessed freedom of movement were animals; those that had the gift of photosynthesis were plants. In 1862, the German biologist Ernst Haeckel proposed that the microorganisms that he called Protista were deserving of the same status as plants and animals, though it wasn't until the 1930s that it was widely accepted as a third Kingdom.

But promotion to the same status as animals and plants left the taxonomy of one-celled creatures at an oversimple level. Ever since Darwin, the informing idea of taxonomic categories such as genera, families, phyla, and so on, is that members should have a common ancestor, rather like a family tree: First cousins share grandparents; second cousins great-grandparents. But the purely morphological technique that grouped like with like—mammals have fur, insects chitin—was too blunt a filter for unicellular life. This is because the most basic difference between living things is not between vertebrates or invertebrates, or even plants and animals, but between those that possess and those that lack a cellular nucleus, the very kernel of life. The Greek word for kernel is *karys*, so the word used to describe creatures without a nucleus is *prokaryotic*—the name, originally *procariotique*, was coined in the 1930s by Edouard Chatton, a French marine biologist—while creatures with nuclei are *eukaryotic*. Bacteria are prokaryotes. Pretty much everything else, from yeast to elephants, are eukaryotes. This realization resulted in the creation of a fifth Kingdom, dividing one-celled eukaryotes, who retained the Protist name, from prokaryotes. Thus, by the time the dust had settled, in the 1970s, the hierarchical tree of life had two domains—Prokarya and Eucarya—and five kingdoms: Plantae, Animalae, Fungi, Protista, and Bacteria.*

A proper recognition of the status of bacteria, however, requires more. Long before it had an underlying theory like natural selection or tools like biochemistry, life science became systematic by categorizing organisms according to morphological differences—legs versus wings, scales versus skin. The bias survives, and profoundly understates the variability exhibited by bacteria. The one-celled creatures have a decidedly smaller range of body shapes than, for example, a single family of insects, but that is because bacteria do not usually evolve to fit evolutionary niches by changing their physical shapes but by their chemistry. To understand the difference, consider the challenge of providing for scientists at an arctic weather station: Instead of hiring a mechanical engineer to construct a heating and ventilation system, you would formulate a pill that changes internal body temperature to cope with subzero temperatures. Bacteria are lousy mechanics, but are chemists *par excellence,* inventors of virtually every metabolic reaction known to science, from respiration to photosynthesis to digestion.[4]

Not all their inventions, however, were so constructive.

———

One defining characteristic of all forms of life is the collection of genetic material that can be combined in new and interesting ways. In the relatively brief two million years that humans have been recombining their DNA, a hundred billion unique versions of the species

———

* Or not. In 1977, a microbiologist named Carl Woese, then at Yale, announced the discovery of a third domain, the lineage of microorganisms living in extreme environments, such as the superheated vents on the ocean's floor where pressure is so great that water stays liquid even at temperatures above 300 degrees Fahrenheit. Since the conditions in which these life forms survive today are believed to mimic the conditions of life on the young and oxygen-free Earth, Woese named them Archaea. Subsequent discoveries have revealed that Archaea—sometimes called Archaebacteria—are actually more closely related to eukaryotes than they are to Bacteria, but since both are studied by bacteriologists, they will be collectively referred to here as small-b bacteria.

have walked the planet. Bacteria have not only been playing the game several thousand times longer than their descendants, they are also several times more resourceful in acquiring genetic material from preceding generations. Three times more resourceful, to be precise, since bacteria possess the ability to obtain genes in three different ways. Bacterial genes are acquired, first and most familiarly, by *conjugation*, cell-to-cell contact by means of a hairlike sex organ called a pilus. In addition, they engage in *transduction*, in which the pilus is replaced by a virus, itself a bit of DNA in a coat of protein, that can carry genes between bacteria. And, since the voracious bacterial appetite for new genetic material is unsatisfied with "only" two methods of acquisition, many bacteria also gather plasmids—DNA floating loose in the bacterial environment—a process called *transformation*.

The impact of this many-pronged reproductive strategy is profound. Virtually all other species of life share genetic material only with members of the same species—occasionally closely related species, as when horses are mated with donkeys. But since a bacterium—*E. coli*, for example—collects both free DNA and DNA carried by viruses, the "breeding" population from which potentially useful code may be selected is not just other *E. coli*. The library of code open to perusal by an unsatisfied bacterial cell is essentially every other bacterium on the planet, as well as a fair share of its viruses. As a result, in the formulation of the great evolutionary biologist Ernst Mayr, who defined species as groups that are reproductively isolated, bacteria are not species at all.

One might think that conjugation, transduction, and transformation, which are collectively called "horizontal evolution," offer bacteria quite sufficient advantages over other forms of life when it comes to creating new and improved species. If so, one would be mistaken, for bacteria—like other organisms—have another method: Evolution occurs vertically as well, through the preservation of helpful mutations.

Mutations are essentially typographical errors in the genetic code, frequently caused by environmental traumas such as radiation. Most mutations, whether they occur in bacteria or humans, are either neutral or injurious, of course; typos rarely add information to any document. But if the document is long enough, one or more of those typos might actually improve an author's work, and even a relatively short work with a high percentage of errors may still come up with a few *bon mots*. A large enough population with a high enough mutation rate will also produce some happy accidents—and even a tiny percentage of a huge number will be itself large in absolute terms.

In bacteria, both population size and reproduction rate are frighteningly high. Mutation rates for any species are calculated by finding the population size at which the chance of a mutation is better than even after a single generation, and for most bacteria that population size is only a little greater than one hundred. Since a cup of seawater can contain literally billions of bacteria that can reproduce a new generation every thirty minutes or so, the potential for a useful mutation approaches certainty every twenty-four hours.

Useful, that is, for the bacterium.

Some time before Justinian completed his journey from his Balkan home to Constantinople's throne—possibly thousands of years before— *Y. pseudotuberculosis* had reshuffled its DNA so completely that it was, by any standard, a new organism. The *how* was some combination of vertical and horizontal evolution; the *why* is a little trickier. So long as the species was restricted to digestive systems, its ability to propagate, the most basic of all biological imperatives, greater even than the instinct to survive, was constrained by its capacity for travel, usually through the relatively slow and unreliable distribution of animal waste. If bacteria could dream as obsessively about propagation as Justinian did about reunification, this particular one would surely have imagined a superior method of transportation from one host to another . . . in the language of pathology, a *vector*, the vehicle that transfers a pathogen between organisms. No one can know how many false starts it took before the bacterium happened upon the ideal vehicle. And then, one day, it did.

The flea.

Though fleas evolved at least sixty-five million years ago,* the first

---

* Fossils resembling modern fleas have been found in sediment two hundred million years old.

man to study them carefully was, in one of history's lesser ironies, the same Anton van Leeuwenhoek who discovered bacterial life. When Leeuwenhoek wrote, "This minute and despised creature . . . [was] endowed with as great perfection as any large animal,"[5] he was understating. The tiny insects—the smallest are scarcely 0.01 millimeter in length—are capable of behaviors that put any flea circus to shame. The European rabbit flea is so synchronized with its host that a female flea only gestates when living on a pregnant doe; when the doe gives birth, so does the flea, and the new fleas find a happy home on the baby rabbit.

With legs powerful enough to propel the wingless insects up to two feet, and hooks at the end of those legs that the flea uses to latch onto host animals, the flea was clearly mobile enough to serve as transportation from one host to another. But to a bacterium, the voyage from host to host is only half the journey; the other half, arguably the more important one, is the passage from the outside of the host to its interior. In fact, as vectors go, fleas are fairly sedentary. Tsetse flies and mosquitoes are extremely mobile carriers of the organisms that cause sleeping sickness and malaria; but a flea, especially one that favors rodents, is likely to spend its entire life on a single animal. As a result, what really mattered to the bacterium was not the vector's legs, but its mouth. The ability to penetrate the skin of host animals in a way that carries bacteria into the animal's interior is the *sine qua non* for a successful vector, and it is the characteristic for which the flea was superbly evolved. For it is the flea's diet, more than its mobility, that makes it uniquely attractive to *Y. pseudotuberculosis:* It is a blood sucker, with two lancetlike darts, called *maxillary laciniae,* on either side of its mouth, that connect the insect's two pumping systems directly to reservoirs of mammalian blood.

It takes a fair bit of nipping and tucking to transform *Y. pseudotuberculosis* into an acceptable flea-borne passenger. In bacteria, since form and function are essentially the same thing, the formation of a new and improved organism means the acquisition of a spectrum of new functions, and that means rewriting genetic code. The bacterium that evolved to live in a mammal's digestive system needed genes that would enable it to live in the very different environment of a flea's midgut, which produces a substance that compresses the rod-shaped bacteria into spheres, and then consumes them. At some point in its rebuilding project, *Y. pseudotuberculosis* acquired DNA code with the instructions for making a protein called YMT (for *Yersinia* murine toxin) that turns off the flea's defenses.[6] But, since the midgut of the flea is only a stopping-off place for the bacterium, it needs a way into the bloodstream of the flea's dinner, and for that, it must also turn off genes that make proteins that adhere to the lining of the flea's stomach, or which are so toxic to the flea that it would be poisoned before delivering its cargo of bacteria. Finally, to be a truly efficient traveler, the bacterium needed to exit the flea's digestive system as successfully as it got there. The exit point for the bacterium is the flea's foregut, a sort of bulb whose interior is lined with seven rows of interlocking spines that control the flow of blood into the flea's stomach. With the acquisition of another bit of code for building a protein called HMS (for *hem*in *s*torage locus) the bacteria taught itself how to make a sort of glue—a biofilm, like dental plaque[7]—out of blood. Using the biofilm, the bacteria were able to clump together, sealing off the flea's stomach from its meal, which not only caused the flea to feed frantically trying to forestall starving to death, but turned the foregut valve inside out, forcing out the blood already *inside* the flea.[8]

William James famously wrote that there is very little difference between one man and another, but that the difference that exists is very important. The difference between *Y. pseudotuberculosis* and its successor organism is very small indeed. Or, more precisely, successor organisms. *Y. pestis* comes in more than one form, though no general agreement has yet been reached on how many. The French biologist R. Devignat, somewhat arbitrarily, distinguishes between three different, and subtle, variations of the *Y. pestis* genome. The first was *Antiqua* (from the Justinianic Plague), *Medievalis* (which is associated with the plagues of the fourteenth through the sixteenth centuries), and *Orientalis* (from the third, largely Chinese, pandemic). Even though no really reliable dating can be made for these three strains—and, as below, even their geographic homes are somewhat obscure—the taxonomy is widely used even today. It is based not merely on historical affinity with the three pandemics, but biochemical properties—one can ferment glycerol, another can turn nitrate into nitrite, one can do both (to compound confusion, a fourth variety, called *Microtus*, can neither ferment nitrate nor cause disease in animals larger than the rodents in which it is chronic). Within the three biological variations, or biovars, at least thirty-five different strains, or lineages, have been identified; a team led by the French microbiologist Michel Drancourt argued persuasively that their studies of the dental pulp in sixty skeletons exhumed from a sixth-century mass grave near Sens, in southern France, indicate that the demon that departed Pelusium in 540 was from the *Orientalis* biovar.[9] Scientists who worked in the Soviet Union bioweapons program have popularized a different scheme, sorting the strains by host mammal (and flea): marmot plague, gerbil plague, rat plague, and ground squirrel plague.

Even more startling than the protean character of *Y. pestis,* or its strong resemblance to its parent organism, with which it shares 95 percent of its genetic code, only a tiny fraction of the new organism's genetic material is new: 32 genes and 2 plasmids.[10] The most telling difference between the demon and its relatively innocuous parent is negative: 317 of *Y. pseudotuberculosis* genes are *absent* from *Y. pestis,* deleted by sections of the bacterial genome—the "insertion elements"—whose function is to scramble the genetic sequence. In only a few hundred years, transposing a few hundred genes, losing a few hundred, and adding a few dozen, *Y. pseudotuberculosis* had given birth to an entirely new species. Yet another random set of genetic bets had paid off with a jackpot; in the words of complexity theorist Stuart Kauffman, a self-maintaining system, "daughter of chance and number, swarmed into existence."[11]

Y. pestis could now travel anywhere that a flea could carry it. But which flea? Dozens of flea species have been shown to be vectors for *Y. pestis,* but the first, and most successful, was the flea known by the

Linnaean name *Xenopsylla cheopsis.** *X. cheopsis* is, like most fleas, a prodigious breeder, laying nearly four hundred eggs over the course of its lifetime. Moreover, though that lifetime is relatively brief, normally cycling through egg, larva, pupa, and imago in two weeks' time, if conditions require, it can survive, in its pupa stage, for up to a year in places like middens and old nests. More important, *X. cheopsis* becomes blocked by bacterial colonies faster than any other flea species—less than five days. From the standpoint of the bacterium, this characteristic is a good-news/bad-news story; the starvation that turns *X. cheopsis* into a manic biter of anything that moves also results in a dramatic loss of life expectancy. Resourceful as ever, *Y. pestis* may have turned even this to its advantage. Sixty years ago, the biologist A. L. Burroughs postulated that one of the reasons that the bacterium resides in dozens of different fleas is because those fleas that are less likely to be blocked by bacterial colonies can preserve populations of *Y. pestis* for long periods;[12] some flea species can survive for more than a year after being infected, thus acting as a strategic reservoir for the bacterium and its genes.[13]

———————

Genes are so central to life as we know it that some evolutionary biologists, Richard Dawkins most prominently, are sometimes quoted as saying that organisms are nothing more than vehicles used by genes to assure their own propagation. But they're not everything. Reordering of DNA is necessary, but it is not sufficient. The other minimal requirement for governing a network of actions and reactions, whether chemical or human, is some border between self and everything else. Life needs walls.

Bacteria are defined by their walls. Literally: A bacterium whose

———————

* Kenneth Gage of the Centers for Disease Control is one of the best-known advocates for a role of *Pulex irritans,* the flea that favors humans, as a carrier of plague. Though this position is popular in Russia, it remains controversial everywhere else, largely because of the disease's rarity; if *P. irritans* were a generally important carrier, humans would be a far richer reservoir for the disease. Moreover, though virtually all of what might be called prescientific epidemiology can only be seen through a fog of anecdote, the patterns that appear through that fog do not do much for a belief in the importance of *P. irritans,* such as the death rates that are randomly assorted among large and small families. If *P. irritans* were the chief carrier, one would certainly expect higher death rates in larger families.

membrane turns purple and stays that way when a crystal violet stain is applied is called gram-positive, while those that turn pink when a different solution is added are gram-negative. The term* is more than a naming convenience. One of the earliest evolutionary advances of bacteria was the creation of a thick cell wall of carbohydrate polymers—a fancy name for sugars—connected via proteins, effectively forming one giant molecule called peptidoglycan. Much later, different species of bacteria wrapped the peptidoglycan in a combination of fat, protein, and carbohydrates. The combination not only resists the purple stain, but gives gram-negative bacteria a sturdier defense against attacks aimed at the sugar-protein wall of peptidoglycan.

The membrane is more than protection, more even than a necessary boundary marker between self and environment. The initial function of the bacterial membrane was probably nothing more than regulating the movement of energy into the bacterial cell. From that relatively simple beginning to the bacteria of today is a difference far greater than that between a lean-to and a cathedral. The bacteria of today build walls of remarkable complexity in order to sustain dozens of ongoing chemical reactions. Each of those reactions obliges the walls to allow some of the environment's molecules to enter, and some of the cell's to escape. To control that movement, the walls incorporate a dynamic system of aqueducts and pumps to move nutrients in and out of the bacterial cell; some of the pumps use a charged molecule to push or pull molecules with an opposite charge into the cell. Other systems use proteins to open channels in the cell wall that permit the entrance or exit of only one material. Still others assemble a train of proteins that pass an amino acid from one to the other, while other proteins hold open the membrane channel.

Even though the walls of the earliest bacteria were far simpler than those of today, the size of the bacteria that they contained has been virtually unchanged for more than three billion years. This is because bacterial size—about one to two micrometers in diameter, one three-

---

* It has nothing to do with the metric measure, but is named for Hans Christian Gram, the Danish physician who discovered the phenomenon in 1884.

hundredth the width of the period at the end of this sentence—is a function of a mathematical cube-square law. As noted above, because the chemical reactions basic to life require the movement of molecules from the outside of a membrane to the cellular material inside, a high surface-to-volume ratio facilitates them. In a cellular factory, no less than in a metal shop, the closer the loading dock is to the fabricating machines, the more shipments can leave the factory.

And there are a *lot* of factories. After more than three billion years of expanding into every possible niche in earth's ecosystem—so long as water is liquid (and even the water superheated to many hundreds of degrees by vents in the ocean floor stays liquid because of the enormous pressure) bacteria thrive[14]—prokaryotes are not only earth's most abundant type of life, but its largest in terms of mass; in fact, for every ten pounds of sea life, nine are prokaryotic. It is scarcely surprising, therefore, with an aggregate size that dwarfs every other life form on earth, that bacteria have the planet's largest and most undiscriminating appetite. Bacteria "eat" inorganic chemicals, sunlight, other bacteria, and every creature that walks, swims, flies, or crawls. But only during the most recent, tiniest fraction of their history have bacterial genes combined in a way that regards animals as food.

———

At the most basic level, all living creatures on earth eat the same food: the compound adenosine triphosphate, or ATP, which stores energy in the chemical bond that is the difference between adenosine diphosphate—ADP—and ATP. Like a battery charged by a flywheel, every time a molecule of ADP is converted to a molecule of ATP, it "costs" eight kcal, but the energy doesn't disappear. It remains available for release as energy when the cycle is reversed. This circular journey, the linchpin of the three processes by which sugars are turned into energy, is such a remarkable energy motor that it is still very much the gold standard even among Johnny-come-lately species like *Homo sapiens.* It is now generally known as the tricarboxylic (or citric) acid cycle, but when the refugee German chemist Hans Adolf Krebs described, in 1937, the dozen steps that convert acetic and oxaloacetic acid into citric acid, forming those high-energy phosphate bonds, he gave it his own name. The intricacies of the Krebs cycle have been torturing secondary school students ever since.

ATP can be produced in two different ways. Organisms that turn sunlight into ATP—that use the green pigment called chlorophyll to absorb the energy contained in a single photon of light and so convert the chlorophyll into a molecule of adenosine triphosphate—are *phototrophic;* everything else is *chemotrophic.* Modern chemotrophes, including humans, produce ATP by consuming both organisms that photosynthesize and other chemotrophes: salads and hamburgers. Simple as this seems, it is not the most basic of life's eating plans. Earth's earliest life forms predate photosynthesis by at least hundreds of millions of years. The very first eating strategy for which experimental evidence exists was simple oxidation: liberating high-energy electrons from elements like sulfur, nitrogen, and hydrogen. One step up the food chain, bacteria metabolized—fermented—the organic compounds that occurred naturally on the early earth. The technique of fermenting sugar, releasing energy that can be used for living and growing, and leaving behind a less-energy rich acid, has proved popular among bacteria to this day; the lactic acid that turns milk sour is the residue of bacteria that gorged on the milk sugar called lactose; so is the alcohol that bacteria leave after eating the fructose in grapes.

However, the sugars available from inorganic carbon molecules are a use-it-and-lose-it resource, one that declines precipitously wherever it is used for food. This left early bacteria with a looming problem of declining resources, in the face of which some particularly inventive bacteria developed photosynthesis, combining the sun's energy with atmospheric hydrogen. Instead of eating ATP "raw" they could use it to fuel an engine that would make sugar out of carbon dioxide and hydrogen, which is as close to a limitless resource as can be imagined. For an immense span of time, even the most voracious bacteria were satisfied by it.

Bacteria owned evolution on earth for two billion years before they finally had some company. Another billion years afer the first eukaryotes had formed, very likely when bacteria learned how to swallow other bacteria, and to convert them them into specialized energy generators like chloroplasts in eukaryotic plants, and mitochondria in animals,* evolution expanded the bill of fare. At a moment in history that is practically last week by bacterial standards, somewhere between 570 to 800 million years ago, life on earth got complicated. An enduring image from classical mythology is the tale of the Titans, earth's first rulers, whose king, Cronos, ate his own children. Like Cronos, Earth's first life form was about to evolve its future larder.

That planet that they occupied had been utterly made over by the activities of bacterial and single-celled eukaryotic life. The photosynthetic† cyanobacteria—in older texts, they are called blue-green algae even though they are not algae, and are usually neither blue nor green—had unleashed on the defenseless earth one of the most corrosive substances imaginable: free oxygen. The "anaerobic" bacteria that had, until then, overwhelmingly comprised the planet's population, faced with extinction, were forever consigned to the remaining habitats without free oxygen, such as the ocean floor. Meanwhile, other bacteria had fixed nitrogen in the soil of the planet's surface, permitting the growth of still more photosynthesizing plants.

In other respects, Earth was unrecognizable to anyone familiar with the modern globe, transformed by the planet's other great engine of change. Plate tectonics is a process that makes natural selection look like a video on fast forward; moving at a rate of only a few inches a year, the enormous continental plates took a hundred million years to complete a trip around the globe. Six hundred million years ago, virtually all the plates were part of one large land mass clustered around the South Pole. Modern geologists call the supercontinent Rodinia,

---

* The great evolutionary theorist Lynn Margulis was the first to argue that utterly critical features of both plant and animal cells—chloroplasts in the former, mitochondria, the organelles in human cells that serve as energy factories, in the latter—are really bacterial cells that have been symbiotically incorporated into larger cells for the last billion and a half years. Though the evolutionist John Maynard Smith calls mitochondria "encapsulated slaves," the theory that eukaryotes derived from prokaryotes is not without critics. C. G. Kurland of the University of Uppsala argues that eukaryotes and prokaryotes share a common ancestor . . . are separate branches off a not-yet-discovered tree.

† The first bacteria to photosynthesize were anoxygenic; that is, they did not produce oxygen as a by-product of metabolism. Only with the cyanobacteria (and later, the plants) did earth's oxygen atmosphere form.

and the sea that covered the remainder of the planet the Panthallassic Ocean—together the site of what would be called the Cambrian Explosion, the introduction of multicellular organisms to the world.

Explosion is the only word for it; not only did the number of phyla increase from fewer than half a dozen to nearly forty, but the planet's total biomass increased to essentially the same level that it is today.[15] For three hundred and fifty million years, living organisms have essentially been swapping the same trillion tons or so of weight between them.*

The most widely accepted theories about the moment when protozoans—single-celled eukaryotes—became metazoans, or multicelled organisms, assume that colonies of the former proved to be such an adaptive improvement that, over time, they became not colonies of independent cells, but separate organisms. The first metazoans, therefore, were probably very like sponges or amoebae, which are essentially groups of relatively similar cells, whose reason for assembling together was the vastly improved ability to eat their own single-celled ancestors. Like the eukaryotes before them, the metazoans could grow even larger and more diversified, and so expand into empty, and therefore attractive ecological niches. The temptation to compare such assemblages to the growth of human societies has proved irresistible to generations of natural historians, and they have a point: Families are at a disadvantage compared to tribes, tribes to kingdoms, and kingdoms to empires.

With a slight shift of perspective, however, the explosion of evolutionary diversity reflected in all those multicellular plants, insects, fish, and vertebrates is not simply an advantage for the metazoans, but for the bacteria: specifically, for the bacterial diet. From that angle, humanity's food chain, and humanity itself, are nothing more than bacteria's way of distilling food into denser and more efficient packages, just as a steak is a concentrated form of the solar energy found in grass and consumed by cattle.

---

* Or perhaps not; a currently fashionable theory is that the mass of anaerobic bacteria living below the earth's surface at very high temperatures matches or exceeds the total surface biomass.

———

The number of bacterial generations that separates the first photo-synthesizers from the Pelusium corpses is in the range of 10 to the 11th power, roughly six orders of magnitude more than the number of generations separating *H. sapiens* from our oldest eukaryotic ancestors. This almost unimaginably long time has produced a frighteningly sophisticated armory of survival weaponry, such as armor that permits bacteria to thrive in subzero glaciers, in vents at the bottom of the oceans where pressure is so high that the water is nearly five hundred degrees, in environments as corrosive as superconcentrated sulphuric acid, and—even more hostile—the digestive system of *X. cheopsis*.

The choice of that particular digestive system was a consequential one. Fleas are overwhelmingly species specific; cat fleas *(Ctenocephalides felis)* only jump to dogs *(C. canis)* if a cat cannot be found. Even more finicky species include *Oropsylla montana* and *Hoplosyllus anomalus*, ever faithful to the rock squirrel, or *Pulex irritans*, the loyal companion to a thousand generations of human beings. From the standpoint of the bacterium, the ideal flea was the one with the ideal host: one with population densities high enough, and a breeding cycle rapid enough, to offer a large and consistent inventory of mint-condition real estate.

The preferred host for *X. cheopsis* is the rat.

CHAPTER EIGHT

"From So Simple a Beginning"

There is grandeur in this view of life, with its several powers, having been originally breathed into a few forms or into one; and that whilst this planet has gone cycling on according to the fixed law of gravity, from so simple a beginning endless forms most beautiful and most wonderful have been, and are being, evolved.

—Charles Darwin, *The Origin of Species*

I T IS A PERVERSITY of language that one of the deadliest diseases in mankind's history—so feared that it has become virtually a metaphor for epidemic illness—isn't even directly contagious between humans, at least in its most common form. Bubonic plague is a *zoonosis,* a disease that makes its home in a population of animals, sometimes marmots, or prairie dogs, or gerbils, where it remains chronic to this day.* But the historical importance of every other plague carrier combined is dwarfed by the impact of the rat.

Europe is home to two different rat species. Today, the brown rat, *Rattus norvegicus,* is by far the most numerous and the dominant species of rat. In the port cities of the sixth-century Mediterranean littoral, only rarely traveling as far as a mile inland, it was *R. rattus*: the Mediterranean black rat.[1] It had been so ever since it arrived from southwest India (where it first shows up in the fossil record), probably on the ships that carried black pepper to the empire from the great entrepôt of Goa. The pepper ships certainly carried this sedentary creature as far as the Horn of Africa, and from there they could have been

* In 1910, sixty thousand Manchurian trappers caught bubonic plague from marmot skins, as reported by Wendy Orent in her book *Plague.*

carried by sea north to the Mediterranean via the canal connecting the Red Sea with the Nile built, in successive pieces, by Darius, Ptolemy, and Trajan.[2]

However the black rat arrived, archaeologists have found evidence of its presence during Roman times throughout Europe. In Corsica, barn owls have left evidence of their taste for rat babies in middens all over the island, dating from the Roman conquest during the Punic Wars. Rat bones from the first century have been found near Amsterdam, in England, on the Rhine, throughout Italy, Spain, France— everywhere that the legions carried Rome's eagles, they also carried her rats.[3]

Rats are not travelers by choice. The two-hundred-meter limit that generally describes the farthest a normal rat journeys during a single life span (typically two years or so) means that, on their own, the eight-inch-long rats can spread less than fifteen miles in a century. So long as there are people, however, rats don't have to travel on their own, and as a result, they live, literally, everywhere humans do. And they do so in densities that make Hong Kong look like the Australian Outback: During an explosion of the rat population on a single Iowa farm, densities of more than one thousand per acre were reported, and rat populations in East Africa regularly exceed eight hundred per acre.[4] This is not the sort of population growth that occurs among fastidious eaters, and rats are as omnivorous as goats. They eat any sort of vegetable matter, from the obvious—seeds, nuts, leaves, and fruits—to such unlikely choices as paper, soap, and beeswax,[5] as well as eggs, baby chickens, pigs, even lambs. But for those charting the rat's impact on human history, their most significant dietary choice is grain. Grain is what induces rats to travel at all. A rat eats fifty pounds of grain annually, and spoils twice as much as it eats. The grain fleets that sailed from Egypt to Rome and Constantinople, and the carts that carried the grain to Roman colonies, were therefore simultaneously the pillars supporting the entire imperial structure, and the path of the transmission for a deadly pathogen.[6]

Movement by sea is easy enough to imagine. The spread of rat-borne diseases, however, was wider than what could simply have been reached by Rome's merchant vessels, though it is still "mappable" with a decent amount of common sense. Rats travel better overland in carts than on horseback, and sure enough, the sometimes eccentric pattern of rat appearances, particularly in Gaul, are closely associated with those Roman roads paved well enough to accommodate carts.[7] The historian Michael McCormick was not overstating the case when he wrote "the diffusion of the rat across Europe looks increasingly like an integral part of the Roman conquest."[8]

Even so, the attractiveness of the rat as a bacterial host is less a matter of its range than of its population density. Bacteria prefer—in the evolutionary sense—a host population in which it can spread most efficiently from individual to individual, and so sustain itself. This, in turn, is a function of the number of rats living within the normal traveling range of individuals—generally about an eighth of a mile—and the percentage of that number that is highly susceptible to the disease. Researchers generally agree that the magic number for a self-sustaining bacterial round robin is six thousand rats per square kilometer: about twenty-five rats per acre, a modest enough achievement for *R. rattus.* Once that population density is achieved, arithmetic takes over: Females are in season at all times, delivering litters of up to twenty pups three to seven times a year, each of whom is capable of breeding on its own less than three months after birth,[9] which gives a flea-borne bacterium like *Y. pestis* regular access to healthy hosts to replace dying ones.

Which is why, in the end, the bacterium became a pathogen. The sedentary character of the rat flea, matched with the sedentary lifestyle of the rat, practically forces *Y. pestis* to become a killer; since the flea depends on the warm blood of a live rat, it has no reason to jump to another rat unless the host rat dies.[10] In fact, the demon is even more diabolical in its selective adaptation, since it requires that its rat host stay alive long enough for the bacteria to reproduce before the rat dies and the bacteria departs. *Y. pestis* exhibits the coordination of a juggler in maintaining the balance between life and death, manufacturing a protein that makes one of its poisons—the technical name is Lipid A—less poisonous at mammalian body temperatures than at room temperature. As a result, rats are asymptomatic for the first five to six days of infection, and only during the latter stages of the disease has the bacteria population grown so large that it blocks the rat's biliary duct, causing massive swelling, and finally death.

The rat's death is the bacterium's life. From that perspective, bubonic plague in *H. sapiens* is nothing more than a "sideshow," which is the word used to describe it by Julian Parkhill of Britain's Sanger Centre, who completed sequencing the genome of the bacterium in 2001. But it is a uniquely deadly sideshow, for reasons having much to do with human history. From the moment humanity originated in East Africa, human populations in the origin basins of Tanzania and Ethiopia grew far more slowly than they would have anywhere else, because they were surrounded by the richest menagerie of pathological microorganisms on the planet, the evolutionary equivalent of a baby crib filled with deadly stuffed animals.[11]

These hazards faced by early humans tend to fall into two categories. The first, the great parasitic epidemics of Africa, are mostly carried by worms and other multicellular life, and even though they don't, as a result, cause the human immune system to produce antibodies,* they do have a self-correcting feature. Once the host population drops below a threshold level (not necessarily because of widespread death; the lassitude caused by many African parasites has the effect of reducing the number of transfer opportunities, even if local densities don't change. Put another way, when people stop leaving their homes because they're too tired to walk, they also stop carrying hookworm to their neighbors)[12] the epidemic burns out. The second category, diseases carried by microorganisms that do cause an antigenic response—most bacteria and viruses—are subject to selection pressures that push them to achieve a modus vivendi with host organisms. The best-known example is the childhood disease of chicken pox, caused by the varicella-zoster virus. Varicella is not only mild enough in its effect on most people that virtually no host dies, but it can remain within the host in a latent form, traveling to tissues of

---

* For more about antigenic responses, see Chapter 9.

the nervous system, reappearing decades later as the disease called shingles.

It is therefore a reasonable rule of thumb when comparing diseases to presume that the more moderate the pathologic effect, the longer the two organisms have lived in contact, mutually evolving. By corollary, those diseases with the least evolved life cycle—those that kill hosts—are newer. They are also far more dependent upon highly dense host populations, which among humans are almost entirely an artifact of the last ten millennia, the period in which agriculture and settled communities started humanity down the path to civilization.[13] Measles, as an example, will die out in a population of less than five hundred thousand, either killing or immunizing everyone before they can produce offspring.[14]

The disease spread by *Y. pestis,* on the other hand, is vulnerable to neither of these evolutionary correctives. Because it is a zoonosis, evolved to survive not in human, but in animal reservoirs, it is virtually impossible to eradicate from a human population. Further, it has no pressure to moderate its virulence so as to live side by side with human hosts. To the bacterium, the only animal that matters is the rat.

To humans, on the other hand, the rat characteristic that matters most is the number and size of local rat populations living near human communities: more rats, more risk. Rat populations fluctuate in direct proportion to two things: availability of food and dry heat. A season with lower temperatures and increased precipitation increases rat populations; one with higher temperature and lower precipitation lowers them. Anything, therefore, that might have caused either an increase in food supply or a decrease in temperature would go a long way toward explaining why it took until 541 for *Y. pestis* to make its presence known in Pelusium. Researchers have documented just such a temperature drop, at precisely the appropriate time.*

While one variable in the risk of plague in humans is an increase in the number of rat populations, another—perversely—is the percentage of rats within those populations susceptible to the disease. Each

---

* For more on sixth-century climate change, see Chapter 9.

rat is home to hundreds if not thousands of fleas, one in eight carrying *Y. pestis* in its foregut, and every time an infected flea bites a rat, it injects up to twenty thousand bacterial cells into the rodent's bloodstream. However, because not all rats in a local population are susceptible to plague, either because of a naturally hardier immune system, or acquired immunity to the disease, susceptibility becomes the key variable in the calculus of the disease. When fewer than half the rats are susceptible, the pressure to make the jump to humans is very low; when the number is greater than 80 percent, the pressure is virtually inevitable. As a result, the calculable "force" of infection oscillates up and down, but never falls below zero.[15]

In short, the greater the number of infected rats, the greater the risk of a population crash, which leaves a large number of hungry fleas intent on biting anything at all. And, while fleas are almost always species specific, starvation makes any animal less finicky about diet. Under such conditions, horses, dogs, even camels become targets for *X. cheopsis*, but the most easily available warm-blooded meal in the flea's universe is the species living in closest proximity to rats: *H. sapiens*. The result is a human epidemic.

The physicians of early antiquity recorded a number of plagues before the advent of the Justinianic pandemic. Tetanus was described and diagnosed by Hippocrates in the sixth century B.C.E., as well as mumps, and probably malaria (though he believed it was caused by drinking stagnant water). Thucydides is quite detailed in his description of the symptoms that accompanied the Plague of Athens in 430 B.C.E.—but even so scholars have debated its causative agent for millennia. Since it was clearly infectious between people—the physicians who treated it were decimated—researchers have long suspected that it was not bubonic plague, but did feature "violent heats in the head, and redness and inflammation in the eyes, the inward parts, such as the throat or tongue, becoming bloody, and emitting an unnatural and fetid breath. These symptoms were followed by sneezing and hoarseness, after which the pain soon reached the chest, and produced a hard cough."[16] It was only in January of 2006 that a team announced that they had found, in fossilized dental pulp from the period, evidence of *Salmonella enterica*, the bacterium responsible for typhoid fever.

The Bible, of course, is a veritable catalog of disease, from the single plague of Ashdod to the ten visited on Pharaoh. As described in the Book of Lamentations, Jerusalem was home to disease victims who "went unrecognized in the streets, their faces blacker than soot. Their skin shriveled upon their bones and dry as tinder . . ." during the city's siege by Nebuchadnezzar in the sixth century B.C.E.; the blackened skin was almost certainly caused by the subcutaneous bleeding caused by scurvy in its last stages. Biblical "leprosy" appears so frequently, and so inconsistently, that no one believes it to be the same disease as what a modern physician would diagnose as Hansen's disease. Often as not, the definitions in the Book of Leviticus more closely describe psoriasis or eczema.

In 396 B.C.E., the Carthaginian army that was then encamped before the walls of Syracuse was stricken with a plague featuring dysentery, skin pustules, and other symptoms. The first-century Roman medical chronicler Celsus describes malarial outbreaks in great detail. Rufus of Ephesus describes a local outbreak of what sounds a great deal like bubonic plague, also in the first century. The two greatest episodes of disease during late antiquity, however, were the fifteen-year-long Antonine Plague brought to the Mediterranean in 165 C.E. by legionaires of Marcus Aurelius who had been warring in Mesopotamia, and the so-called Plague of Cyprian of 251–266, named for the Bishop of Carthage. Frustratingly—and despite the presence, during the Antonine Plague, of Galen, the most influential medical writer of all time—neither disease was described well enough for positive identification, though the consensus is that the first was smallpox and the second, measles. The conclusion is partly based on the tragically *well*-documented experience of New World peoples when they were exposed to both diseases without any acquired immunity, and partly on the epidemiological rule of thumb that suggests that modern-day childhood diseases are products of coevolution from far more dangerous versions. And dangerous both were, killing hundreds of thousands, possibly millions, living along Roman and Persian caravan routes,[17] thousands daily in the largest cities, and possibly even precipitating the Crisis of the Third Century.[18]

Whatever their etiology or consequences, the diseases of history (and prehistory) are still obedient to the same mathematical models used by modern epidemiology, which, in turn, depend upon a single quantity: the average number of individuals infected by a carrier during the infectious period. In a simplified version, this quantity, which epidemiologists call the basic reproductive number, or R_0, is the product of three different variables:

1. The rate of contacts, c
2. The duration of infectiveness, d
3. The probability that a pathogen will be transferred from an infective source to a susceptible host, p

Thus, the average number of contacts is the product of c x d, and R_0 is c x d x p.

When one infectious case produces, on average, one new infectious case before it is no longer infectious itself, $R_0 = 1$. Such a disease is endemic, but fundamentally unstable; it takes surprisingly little pressure to push it over the boundary separating it from an epidemic. In *The Tipping Point*, Malcolm Gladwell famously imagined an untreatable strain of flu carried into New York by one thousand Canadians. He set the transmission probability for this particular strain at a hypothetical 2 percent, and the rate of contact at an (equally hypothetical) fifty per day; he then simplified the example by making the duration of infectiveness identical with the disease's symptoms (a "twenty-four hour bug"). In such a case, the disease is in perfect equilibrium: The one thousand Canadians contact fifty thousand New Yorkers every twenty four hours; 2 percent, or one thousand, of them get the flu. Every twenty-four hours, one thousand people get well, and one thousand people get sick. $R_0 = 1$. But increase the rate of contact by only five per person—from fifty to fifty-five daily, $R_0 > 1$ . . . and an equilibrium becomes an epidemic, with more cases every day than the day before, continuing until the pool of susceptibles eventually declines, either because of acquired immunity or death.[19]

Any of the variables used to calculate R_0 has precisely the same impact. Anything that increases the infectious period or the rate of contacts can turn an endemic disease into an epidemic. In the case of a disease with two intermediate vectors, like bubonic plague, the rate of contact, or c, is the product of the varying sizes of the flea and rat populations, which is to say that even a modest increase, for example, in the rat population can have enormous epidemiological leverage. A worldwide climate change during the late 530s has been extensively documented, and is a good candidate for the precipitating event that pushed a sufficient number of local rat populations over the six-thousand-per-square-kilometer hurdle.[20] The other variable is susceptibility, the probability of transfer from source to host, and it is an even stronger index of the force of infection. The proportion of rats that are susceptible can change for a variety of reasons, but it *always* changes during the times between epizootics. In a year in which a rat population is exposed to *Y. pestis*, some will die from the disease, others live; the surviving population, by definition, contains fewer rats that are susceptible to the pathogen. Contrariwise, in a year without an epizootic, the population at year's end will still contain many more of the susceptible rats. As a result, the percentage of susceptible rats in any given population always increases in the absence of the disease.[21]

Anything that caused the rat population in a region to explode at the same time that susceptibility was on the rise would be explosively bad luck. To a civilization caught in the explosion, the consequences would be catastrophic.

———

A midair collision between two 120-foot-long airplanes traveling hundreds of miles an hour through uncounted millions of cubic miles of space is an unimaginably improbable event. Even so, every year, one or two occur. This is partly because of the sheer number of planes flying, but mostly because the paths they take are not random, but restricted to the far more constrained paths of efficient travel between destinations. In the same manner, the collision between the Roman

Empire and the bubonic plague, despite requiring the coordination of a pathogen, two separate animal vectors, and an intersection in both space and time, was not as statistically unlikely as it appears. The number of locations occupied simultaneously by bacterium, rat, flea, and human, was not only not infinitely large, it was small enough to be counted on the fingers of a single hand.

The scholars and scientists who have spent their lives searching for the birthplace of bubonic plague tend to start their investigations in the present, "walking back the cat" as cleverly as they can. The modern world has literally dozens of basins where the disease is chronic in animal populations,* but only three are thought to be more than a century or so in age, and they match up with Devignat's three variant strains of *Y. pestis*. Given the fact that, of all the players in the drama—bacterium, flea, rat, and human—*Y. pestis* is by far the newest, it seems certain that it evolved in one of those three locales: the Himalayan foothills, the Great Steppe reaching west from China, or the Great Lakes of East Africa.

The demographer J. C. Russell was, for many years, the most passionate advocate of a Great Steppe origin, a view that still has adherents today. Russell's argument is fairly weak, however, given the lack of any real evidence for plague in the extensive prairie that runs from Mongolia to Ukraine until the year 610[22] and not really very much at all until the second great pandemic that devastated Europe in the fourteenth century. Choosing between India and Africa is more difficult; Pauline Allen argued persuasively for an Indian origin, citing the migration of *R. rattus* westward as a stowaway in uncounted thousands of cargo ships sailing back and forth between Ethiopia and India.[23] However, the historian Peter Sarris rightly observes that India is much closer to, and had significantly *greater* contact with China, than with the Mediterranean, yet bubonic plague appeared in China at least sixty years later than it did in Alexandria. Sarris is cautious about

---

* When a zoonotic disease like bubonic plague is chronic in an animal population—neither dying out nor fast spreading—it is called enzootic. When *Y. pestis* resides in human commensal animals—overwhelmingly, rats—the disease it causes is called *murine* plague; when it lives in wild animals, it is *sylvatic* plague.

giving too much weight to this, since the sea routes between Ethiopia and India are far more direct than the overland routes connecting India with China. However, Persia also encountered the plague after Alexandria and Constantinople, and stands squarely astride the land route between the former cities and China. If India were truly the demon's birthplace, it was a rather meandering demon, one that originated in India, spread to Africa and China, and then bypassed Persia on the way to the Mediterranean.[24]

Yet another theory holds that the origins of the plague were in Pharaonic Egypt, when the rat flea *X. cheopsis* jumped from its favored species, *Arvicanthis niloticus,* the Nile rat, to *R. rattus,* the immigrant species from India.[25] Among the odder hypotheses about the origins of the disease is the one that appears in Fred Hoyle and Chandra Wickramasinghe's *Our Place in the Cosmos*, which posits an extraterrestrial birthplace for the plague, whose periodicity is best explained by such activities as sunspots. More persuasive, even though lacking the sophistication of a modern epidemiologist, is the contemporary chronicler Evagrius Scholasticus, who records that the disease "took its rise from Aethiopia, as is now reported, and made a circuit of the whole world in succession. . . ."[26]

(One bit of inferential data supporting an African genesis for the plague is offered by David Keys's examination of the four great sixth-century East African port cities—Opone, Essina, Toniki, and Rhapta, all now vanished—which were the ancient world's largest suppliers of ivory, sending up to fifty tons of the precious stuff, the product of five thousand elephants, up the Nile every year. If the plague started in East Africa, he reasoned, one would certainly see an effect in ivory production, and so he did. From the years 400–540, 120 major ivory artworks [out of an estimated 400,000 made] survive; from 540–700, only 6 survive.[27])

If the demon were born in the fertile African valleys between Lake Tana in the north and Lake Rudolf in the south, lands that are to new species what Iowa is to corn, it would have had its choice of northward routes aboard its flea/rat hosts, either via the Red Sea, or up the Nile, past the point, at Khartoum, where the Blue Nile and White

Nile combine, past the six cataracts that separate Upper Egypt from Lower, and north to Pelusium, Alexandria, and the Mediterranean.

As with much else, the Mediterranean is the key to understanding the unique status of *Y. pestis,* and the disease it carried. Bacterial pathogens had been afflicting humanity for tens of thousands of years before the sixth-century arrival of those bubo-ridden corpses at the mouth of the Nile. But none of them ever swept across what amounted to the entire known world, ending tens of millions of lives, and stopping tens of millions more from ever being born. The reason, simply put, was a mismatch of speed.

Until the arrival of the demon at the mouth of the Nile, at Pelusium, the diseases of civilization had spread at roughly the same pace as civilization itself. Local populations could and did grow, even creating empires of many thousands of square miles—Akkadian, Egyptian, Sumerian, Babylonian, Persian—but the pace at which their merchants and armies traveled was, in general, slow enough to place a brake on the spread of disease. Like a fire, any virulent pathogen would either be tamed, or would consume all available fuel, but it could not sustain itself for long. Overland trade was so slow until early modern times that the only reliable way to spread a new disease faster than coevolutionary adaptation was by water, either an inland sea or rivers.

It was, therefore, probably inevitable that the first pandemic would strike a Mediteranean civilization, rather than a Mesopotamian or Chinese one. The great Belgian historian Henri Pirenne wrote floridly, though accurately, "Of all the features of that wonderful human structure [i.e., the Roman Empire], the most striking, and also the most essential, was its Mediterranean character. . . . In the full sense of the term, *Mare Nostrum* was the vehicle of ideas, and religions, and merchandise."[28] A more mundane look at the enormous advantage in time and money of moving goods by water rather than land—it cost less to ship a bushel of wheat from Palestine to Spain than to send it seventy-five miles by oxcart—suggests not merely the critical character of the inland sea, but also the decisive importance of inland waterways. Egypt was the imperial granary for five centuries not merely because of its farms' productivity, but their location; none of its cultivated land was far from either the Nile or a canal.

In the year 540, at the terminus of the ancient world's greatest riverine complex, the delta of the Nile awaited the arrival of a conqueror greater than Alexander himself.

CHAPTER NINE

"The Fury of the Wrath of God"

540–542

IN 1883, THE Victorian naturalist and writer Thomas Henry Huxley—
"Darwin's bulldog" and the world's first self-declared agnostic—
described Alexandria, as seen from the Mediterranean, in singularly
unimpressive terms. "Nothing can be less attractive than the flat shore
which stretches east and west as far as the eye can reach, without an ele-
vation of more importance than bare and barren sand dunes to break its
even line."[1] The description owes much to the fact that the city built
on that flat shore was, by then, at least a thousand years past its prime.
Founded in 332 B.C.E. by the Macedonian conqueror who gave it his
name, and laid out by Alexander's personal architect, Dinocrates, on the
site of the ancient Egyptian city of Rakotis, Alexandria was intended from
the start to be one of the world's greatest metropolises. After the death of
Alexander nine years later, it became the capital of the great conqueror's
general and viceroy, Ptolemy I, who founded not only the dynasty that
would bear his name but the Mouseion, the great university at which
Euclid, Archimedes, Plotinus, and Ptolemy-the-astronomer all studied.

By the sixth century, Alexandria's reputation for scholarship had
been deteriorating in direct proportion to the stature of Christianity
in the city, which had grown from small beginnings—St. Mark made
his first conversion in Alexandria in 45 C.E.—becoming one of the acknowledged five great bishoprics of the Church. The connection was
not coincidental; over the course of three decades beginning in 391
when Christians destroyed both the Mouseion Library* and Alexan-

---

* The loss of the contents of the great library of Alexandria is the object of a startling number
of myths. The earliest lays the responsibility at the feet of the legions of Julius Caesar; the latest, not only false but libelous, blames the fire on the Islamic invaders of the seventh century.

dria's Serapeum (the "church" for a religious cult built around Serapis, and the center of the Ptolemies' religion) and ending in 415 when a Christian mob killed the Neoplatonist philosopher Hypatia, the city had experienced the dangerous consequences of mixing "secular" scholarship with revelatory religion.

It had not always been thus. During the third century B.C.E., the island of Pharos in Alexandria's harbor—the home of the great lighthouse, one of the Wonders of the Ancient World—was where the Hebrew Bible was first translated into Greek, a direct consequence of the fact that Alexandria was at that time home to the largest Jewish population in the world, and by far the most Hellenized. In legend, at least, the translation was the work of seventy rabbis, confined in seventy different rooms on the island in Alexandria's harbor, who nonetheless produced seventy identical Greek translations of the Hebrew Bible, as well as the Wisdom of Solomon, an anonymous text integrating Greek philosophy with Torah. Those who delight in odd historical connections note that while the Wisdom of Solomon never made it into the Jewish canon, it did bequeath to subsequent generations the personification of Wisdom, an interlocutor between God and man. The anonymous author of the Wisdom of Solomon named his main character Sophia; her greatest monument, the Hagia Sophia, has proved one of the more oblique, though enduring, connections between Constantinople and Alexandria.

It was far from the only one. Christianity might have turned Alexandria away from certain varieties of philosophical speculation, but nothing could challenge its commercial importance. As a sixth-century chronicle observed, "Constantinople . . . is for the most part fed by Alexandria; likewise the eastern regions are supplied, especially on account of the army of the Emperor and the war with the Persians,

---

The durability of the latter legend owes something to the frightening logic of the quote attributed to the Caliph Omar, who was wrongly accused of ordering the destruction of the library on the grounds that its contents were either in opposition to the Quran, in which case they were heretical, or consonant with it, in which case they were redundant. In reality, the reason for the loss of so many of the great works of the classical world—plays by Aeschylus and Aristophanes, works of science by Archimedes—is almost certainly more mundane: deterioration of the parchment, papyrus, and paper on which they were written.

because no other province can suffice for that purpose but the divine Egypt."² Food was not the only, or even the most significant product that Alexandria's ships carried throughout the Mediterranean: "Alexandria possesses one thing which is produced nowhere but Alexandria and her district; without this neither her courts nor private business can be directed; indeed without this very thing the whole race of men could hardly exist. What is this which is so lauded by us? The paper which she makes herself and exports to the whole world, showing this useful thing to all; only Alexandria possesses this above all cities and provinces, but without envy she gives her advantages to others."³

Grain, paper, oil, ivory, slaves. During the year 540, the great merchant fleets of Alexandria, by then officially *ad Aegyptum*, not *in Aegypto*, awaited their next great export, then making their way up the east coast of Africa. The Nile delta was the place where a boom in the population of *R. rattus*, the black rat, reached critical mass.

———

The northward journey of the Nile to Pelusium and Alexandria has a physical inevitability to it, its waters flowing downhill from the mountains of east central Africa to the "flat shore" that Huxley found so unappealing. The northward journey of *X. cheopsis*, the bacterium-carrying rat flea, was a chancier prospect. This is because the flea is only truly active within a very narrow range of temperature: from about 59–68 degrees Fahrenheit, and the drop in altitude that carried the Nile north through Sudan and Egypt also raised the average temperature well past the optimal environment for the bacterium-ridden fleas. On the other hand, the rate at which *Y. pestis* tends to coagulate inside the flea, forming the blockage that drives the flea to bite everything it can reach, tends to fall whenever the temperature goes above 75 degrees. The narrow range bracketed by flea activity and bacterial blocking had kept the disease confined to its East African focus for hundreds if not thousands of years.⁵ It was destined to stay there until something caused temperatures to drop dramatically for a period of years, and then allow them to revert to normal. The timing of such a climate change is therefore extremely important.

The best-known scientific method for establishing the age of objects without historic documentation is probably the measurement of a particular isotope of carbon, which, being radioactive, degrades to a more stable isotope at a very regular rate. Carbon-14 dating has its limits, including its inability to accurately date nonorganic material, but its precision is best suited for *really* old objects, where an error factor of fifty to one hundred years is perfectly acceptable. For accuracy on a timescale measured in thousands of years, rather than thousands of millennia, another method is needed. The technical name for one method is dendrochronology, the measurement of the number and size of tree rings. Numerous species of trees add a ring every year, and in temperate climates—those with distinct seasons—those rings differ in width depending upon the year's climate. Thus, the number of rings in a piece of lumber in a prehistoric structure gives an accurate age for the tree that supplied it on the date it was cut down, and, by comparing the varying widths of those rings to some known baseline temperatures (from historical documentation, other trees, and so on), an accurate date for the tree's harvest, and therefore of the construction date, can be calculated. A tree with three hundred rings was three hundred years old when cut down; if the sequence of the 134th, 135th, and 136th rings matches up with the established temperature of the years 750, 751, and 752 C.E., then the tree must have been harvested in 916.

The process also works in reverse; that is, with a tree of a known date, the temperature of a particular year can be calculated with a surprising degree of precision. Within the last decade an Irish dendrochronologist named Mike Baillie has documented just such a temporary drop in temperature during the mid to late 530s.[6] Evidently, mean summer temperatures, insofar as they can be inferred from the tree rings of Scandinavia, were at least a degree and a half Celsius (3 degrees Fahrenheit) lower than average . . . enough, in fact, to lower the climatic roadblock that might formerly have kept the fleas and their cargo confined to the tropics.

The probable cause for such temporary climate change was a dust-veil: something that blew a huge number of light-blocking particles into the upper atmosphere. Researchers have arrived at a general consensus for the existence of such a dust veil during the late 530s, though not for its source. Baillie favors a cometary shower as the likely cause: "My view is that we had a cometary bombardment—not a full-blown comet, or we would not be here, but parts of a comet."[7] He has contemporaneous history on his side. While one should be skeptical of historical accounts of great comet sightings, which sometimes seem as common as sunrises, both the comets that reportedly accompanied the death of the legendary King Arthur, and the better documented "swordfish" comet that appears in Procopius's history as seen for "more than forty days"[8] occur at an awfully convenient time: Arthur's death is variously dated to 537, 539, and 542, while Procopius dates the appearance of his comet to 539. (Though Procopius also identifies an entire year—536 to 537—during which "the sun gave forth its light without brightness ... and it seemed exceedingly like the sun in eclipse, for the beams it shed were not clear. . . .")[9]

The other candidate for the dust veil is a volcanic eruption, which is the eponymous "catastrophe" referred to in David Keys's book. Keys specifically argues the case for a gigantic eruption from the Sumatran volcano Krakatau, and he is not alone in advocating for a volcanic cause of the climate change. One of the strongest proponents for a volcanic source for the various dust veils is climatologist Richard Stothers from the Goddard Institute for Space Studies, who has mapped each major plague outbreak to a particular volcanic eruption and accompanying broadcast of sulfur dioxide, which turns into sulfuric acid when combined with water. According to this scenario, water vapor in the air thus turns the gas into sunshine-blocking particles of acid, which, once in the jet stream, can be spread across the entire globe in a matter of days. The sulfuric acid aerosol veil has, like the tree rings, left a footprint . . . in this case, a layer of acidic ice in cores dug out of the permanent ice of Greenland and Antactica. Given the distance separating the two ice-core samples, it is safe to assume that the effect was worldwide.[10]

By whatever cause, blocking a portion of the sun's light opened a large enough climatic hole for rat, flea, and bacterium to make their way north to Pelusium, and Alexandria.*

Both advocates of Darwinian selection and their opponents are prone to cite the most benign adaptations as support for their respective positions. Favorite examples of the creationist side—those who style themselves proponents of intelligent design or ID—are much attracted to such complex structures as the human eye or the falcon's wing, adaptations whose intermediate forms seem to confer no real evolutionary advantage, and must therefore have been created *ex nihilo*. The bacterial equivalent, among ID proponents, is the remarkable organic motor that the tiny organisms use as a propulsive device, the flagellum.

One of the leading lights of ID, Michael Behe, regularly cites the flagellum, a collection of more than forty separate parts including a rotor, bearings, and a spiral-shaped "propeller" that hooks onto one of the outer rings. The engine for this propeller is, like everything invented by bacteria, a triumph of physical chemistry: When a positively charged proton moves across the bacterial membrane, the charge causes the ring to turn, which in turn spins the propeller at speeds of up to 15,000 rpm. Because—in the view of ID, anyway—the flagellum's components are only useful as a propulsive device once all pieces are present, it is a challenge to natural selection; a propeller without an engine, or a bearing to turn, is just so much protein. The "irreducible complexity" of the flagellum, goes this theory, is unmistakeable evidence of a designing intelligence.

The weakness—one is tempted to write "irreducible weakness"—of intelligent design is the assumption that the component parts of the flagellum (or the eye, or wing) have always had only a single purpose, a purpose that can only be fulfilled with a complement of other parts. One can only wonder what ID makes of the Type III secretion

---

* Another route, with some vigorous adherents, has the rats traveling north not along the Nile, but as part of the Red Sea trade that carried the ivory artifacts mentioned in Chapter 5 north.

system, a sort of multiple launch artillery system found on the outside of the *Y. pestis* cell membrane, since the same proton pump that drives the flagella in motile bacteria also serves as the delivery system that the bacterium uses to batter down the walls that protect their victims . . . and both pumps share a common ancestor.[11]

That bacteria in general, and *Y. pestis* in particular, have developed such a sophisticated siege engine, using the same parts that can be used to build an outboard motor, is actually testimony to the power of natural selection rather than design. The Type III secretion system is only a part of the armament needed to overcome the enormously so-phisticated defenses produced by a million-year-long arms race be-tween humanity and microorganism. Immunologists conventionally divide those defenses into two systems that operate cooperatively and sequentially: the Innate (or non-Specific) and the Adaptive (or Specific).

The Innate system, first and foremost, is a set of walls built to guard the body's potential points of entry for foreign invaders. The skin itself is part of this system, as are the mucous membranes, the anti-bacterial enzymes in saliva and tears, and the acids of the digestive system that make it difficult for bacteria to adhere to various points in the gastrointestinal tract.

The body's "defense-in-depth"—the first responders to pathogenic invasion that are charged with keeping the invaders contained while alerting more mobile and powerful forces to come and destroy them—is a thing of almost Rube Goldberg–like complexity. It begins with a group of proteins always present in the circulatory system named Toll-like receptors, or TLRs—"Toll" means "weird" in Ger-man, and the name comes from the entomologist who first noticed that fruit flies lacking them tend to have some truly weird bodily dis-tortions: heads on backward, for example.[12] Ten different TLRs have been identified in humans: TLR3, for example, recognizes the genetic material of viruses; while TLR5 recognizes flagellin (the protein used by bacteria to make their whiplike tails).

Once activated, the TLRs, and their surrounding (and still not fully understood) mast cells, summon another set of proteins that have been given the catchall name of cytokines. Each cytokine's individual shape permits it to bind to receptor proteins contained in the invading bacterium like a terrier grabbing a rat. The act of binding sets off a figurative alarm bell, calling the border troops—the circulating antibacterial white blood cells known as leukocytes and phagocytes—to the site of the invasion, causing the infection's first noticeable symptoms: the surface redness caused by increased blood flow; swelling from the arrival of white blood cells; heat from turning up the thermostat located in the hypothalamus (to improve the performance of the white blood cells); and pain from pressure on local nerves. Collectively, the symptoms are known as inflammation.

Sometimes the inflammatory response gets out of control, and the border troops run wild, killing both the patient and the invader; overproduction of the cytokine called Tumor Necrosis Factor-Alpha will push the inflammation over the line into septic shock, or even death. But when it works properly, the white blood cells not only engulf and kill most of the invading bacteria,* but convey them to the lymphatic system, the courier road along which the immune system's field armies can march to decisive battle.

Those field armies, the body's Adaptive Immune System, are composed of specialized cells, the B- and T-lymphocytes, that have "learned" to respond to a particular invasion. The B-lymphocytes produce antibodies whose surfaces contain "hands" that can fasten onto reciprocally shaped molecular handles on the surface of pathogenic cells, while T-cells punch holes in the invasive cell's membrane. Even after the destruction of the invading pathogen, other B- and T-cells remain behind, having memorized the identity of the invaders. The value of these B- and T- memory cells is measured in time; new adaptive cells can take up to two weeks to be "trained" the first time that the body encounters a new variety of invader, but only a day or two

---

* Phagocytosis sometimes involves coating the invader with a substance called opsonins, Greek for "sauce"; once sauced, the invader is more palatable to the phagocyte, which eats it.

once the memory cells have recorded the blueprints needed to produce them again, which is how the body acquires immunity to numerous diseases.

A formidable set of defenses, to be sure. Not only are the body's vital organs walled off from the sea of bacteria in which they swim, but in the event of invasion, can prevent the invaders from finding a place to adhere, can attack with specific and nonspecific arsenals, can even cause the invaders to clump together to make the job of destruction easier. Or so it is supposed to work. The ancient military maxim that "no battle plan survives contact with the enemy" could have been engraved in the genetic code of *Y. pestis.*

The body's first line of defense, of course, proved no obstacle to the bacterium, which had enlisted *X. cheopsis*'s two maxillary laciniae as its battering rams. *Y. pestis*'s real genius, however, reveals itself on the other side of the walls, where the fat-protein-carbohydrate molecule that comprise the bacterium's outer membrane—technically lipopolysaccharide, or LPS—would, under normal circumstances, act on one of the Toll-like receptors in the same way that a red flag acts on a bull. When TLR4 encounters an invader with the LPS coat characteristic of that class of bacteria known as gram-negative, cytokines come running, followed by the inflammation that means the body's defenses have been activated. But run-of-the-mill gram-negative bacteria do not have Yops.

Yops—for *Yersinia* Outer Proteins—are an especially demonic set of compounds produced by the outer membrane of *Y. pestis* that have evolved to counter virtually every defensive measure of the body's immune system. YopJ, for example, prevents the production of cytokines, the couriers that alert the rest of the immune system that it is under attack. Another Yop, YopH, deactivates the proteins used by phagocytes to adhere to the invading bacteria, but still allows the formation of a bubble called a phagosome. The phagocytes then carry the bacteria to the lymphatic system, but without the ability to kill it along the way.[13]

Most Yops work to ensure bacterial survival, but at least one may be a potential host-killer. While its exact role remains uncertain, YopM binds thrombin, which interferes with the body's ability to form platelets, needed for an effective immune response. Simultaneously, one of *Y. pestis*'s plasmids, the free-floating bits of genetic code within the bacterium, contains a set of instructions for something called plasminogen activator protease. As its name suggests, plasminogen activator protease converts plasminogen to plasmin, which digests fibrin, and prevents clotting the blood. The combination of the two disruptions in the bloodstream's clotting mechanism can create both thromboses *and* hemorrhages—the syndrome is called DIC, for disseminated intravascular coagulation—and can cause gangrene, even death. The blackened extremities, chiefly toes and fingers, of late-stage plague victims would give a later pandemic its best-remembered name: the Black Death.

Normally, however, the bacterium has hidden itself well enough within the white blood cells attempting to engulf it that its victim notices no symptoms until the invasion reaches its goal: the lymphatic system, which stretches throughout the entire body. It is the perfect highway for the *Y. pestis* cells, which are reproducing as fast as they can turn human protein into more bacteria. Once inside the lymphatic system, the sheer bulk of bacteria causes the lymph nodes to swell, heat, and become tender. Lymph cells swell enough, in fact, that their own cell walls rupture, and the fluid forms an edema in the lymphatic cells. The result is a bubo. Sometimes the bubonic edema turns septic; like a snake shedding skin, first the lymphatic tissue sloughs cells, then the surrounding muscle, until it reaches bone.[14]

But even the attack on the lymphatic system is only a skirmish compared to the assault to come. The point at the end of the very long bacterial spear is *Y. pestis*'s complement of toxins, the most potent poisons known.

The toxins produced by bacteria fall into one of two categories. The first, called exotoxins, are the more virulent of the two, and can kill a two-hundred-pound human in doses as low as one ten-millionth of a gram. The existence of such lethal substances is a challenge both to believers in an intelligent designer and to Darwinians convinced that every expressed behavior is an adaptation of some sort. There appears to be no reproductive advantage to toxin-producing bacteria . . . quite the opposite, in fact, since the lives of most bacteria end with the death of their host. Many, if not most, such bacteria do not even require a host, and—perhaps more to the point—when living inside human hosts frequently take on a form identical to that of their toxin-producing cousins but still lack the ability to produce it.[15] In the words of William McNeill, wondering why (in evolutionary terms) bacteria produce toxins is pretty much like asking why they also produce therapeutic medicines; in neither case can one persuasively see the hand of either God or natural selection.

But while no one can, with any confidence, offer a reason for the existence of exotoxins, huge amounts are known about their structure, their activity, and their extraordinary specificity. The toxin of *C. tetanus*, for example, attacks only the nervous system, destroying the synapses that inhibit muscular contractions, and thereby cause spastic paralysis; cholera enterotoxin stimulates an enzyme in the GI tract that causes water to secrete, killing by dehydration; staph toxins excite the immune system's inflammatory response, causing vomiting and blisteringly high fevers.

Exotoxins, powerful though they are, produce antigens and can therefore be defeated by the body's immune system. Less virulent, but more stable, is the other large category of bacterial poisons: the endotoxins.

Y. pestis gets its deadliness from its endotoxins, which are produced by the disintegration of the lipopolysaccharide in the cell wall. The LPS wall is needed not only for all the outer-membrane functions that permit the survival of each individual bacterium, but, like a belt of explosives worn by a suicide bomber, is its most powerful weapon as well . . . in the words of Nobel Prize–winning biologist Peter Medawar, "probably the most complicated chemical substance known."[16] And if the point of the spear for *Y. pestis* is the lipopolysaccharide-produced endotoxin, the very tip of that point are the six (occasionally seven) fatty acids, collectively known as Lipid A.*

---

* Chemistry majors note: The accurate description is: a phosphorylated N-acetylglucosamine (NAG) dimer attached to either 6 or 7 saturated fatty acids, some attached directly to NAG, the rest to other fatty acids.

Lipid A is what really turns a fleabite into bubonic plague. Once the cell wall in which the Lipid A is embedded is destroyed—and if the phagocytes don't destroy it by the time it reaches the lymphatic system, the bacterial cells destroy themselves, helped by a protein released by the Type III secretion system—the fatty acids are released. These bacterial suicide bombers cause overproduction of a number of the immune system's most powerful agents, including the enzymes that release lysozyme, and TNF (tumor necrosis factor). The result is shock, sometimes fatal. The diabolical beauty—there is no other word—of the system is the bacterium's ability to delay release of its cargo of Lipid A until its arrival in the lymphatic system, which can carry it throughout the body. Frequently, the system works so well that by the time the first symptoms—"chills, rigor, high temperature"—appear, the body is saturated in bacteria; the "liver, spleen, and lymph glands are no longer human organs. They are tissues of plague, plague bacteria in almost pure culture."[17]

The bacterium's favored path is the lymphatic system, with its ductways leading to nodes under the arms, in the groin, and in the throat. Sometimes, however, the highway leads to the lungs, in which case the victim starts to produce sputum. Until the sputum starts to accumulate blood, nothing much happens, but within days, bubonic plague turns into pneumonic plague . . . and, as a result, particles too small to be seen can now be spread from one human victim to another, and are so contagious and so virulent that breathing in even a single one results in mortality rates of 100 percent. Between five and thirty-one days after exposure to *Y. pestis,* infection occurs; thereafter, death is only a matter of time, on average seventeen days.

But how many deaths? Along with virtually every other aspect of the plague, the number of its victims is subject to virtually constant revision. As late as the late 1980s, the textbook answer to the deaths produced by the Justinianic Plague was one hundred million, which is at least three times too high. Similarly, in 1969, T. H. Hollingsworth combined state-of-the-art mathematical epidemiology with historical accounts to estimate that Constantinople lost 244,000 people out of a preplague population of 508,000.[18] However, he also assumed that the most prevalent form of the disease in the capital was pneumonic plague, which is considerably more deadly than the bubonic form . . . 100 percent fatal, rather than "only" 40 percent. Depending upon the percentage of victims contracting one form of plague or the other, and on the highly problematic mortality estimates derived from modern visitations of plague—in India in 1896–1917, plague death rates were less than 10 percent[19]—the range of deaths during the critical three months of 542 can be calculated at anywhere from sixty thousand to two hundred thousand. Even the lowest number represents a catastrophic increase over the "normal" death rate, probably no more than two or three dozen a day.*

As the spring of 542 turned into summer, nearly a hundred thousand residents of the imperial capital had only weeks to live.

> This disease always took its start from the coast, and then went up to the interior.[20]
>
> —Procopius, *History of the Wars*, II, xxii, 8

It began, as it always began, at the docks. From there, it climbed inexorably, week after week, a rising tide visiting the judgment of God on His people. Though Procopius could not, of course, know this, the plague was borne not by miasmic air (or by the supernatural creatures that he reported some victims saw before becoming ill), but by the rats that carried their fleas into every neighborhood. The geography of the city conspired with them; one reason that the emperor Constantine was inspired to name his capital New Rome was that, like the original model, it had seven hills. The base of each one was stricken by the demon first, giving those who lived on the hilltops the same useless warning as the occupants of the top deck of a sinking ship.

The best evidence is that the peak of the first wave of the epidemic lasted only four months, after which both rat and human populations had crashed so fast that *Y. pestis* could no longer spread. But for one hundred days, Constantinople was a window onto Hell. Every day,

---

* Consider, as a comparison, that in 2002, the average daily death toll among the 7 million residents of New York City was 159.

one, two, sometimes five thousand of the city's residents—one in one hundred of the preplague population—would become infected. A day's moderate fever would be followed by a week of delirium. Buboes would appear under the arms, in the groin, behind the ears, and grow to the size of melons. Edemas—of blood—infiltrated the nerve endings of the swollen lymphatic glands, causing massive pain. Sometimes the buboes would burst in a shower of the foul-smelling leukocytes called pus. Sometimes the plague would become what a modern epidemiologist would describe as "septicemic"; those victims would die vomiting blood from internal hemorrhages that formed even more rapidly than the buboes. Those who contracted septicemic plague might have been the fortunate ones; though they all died (bubonic plague kills "only" four to seven out of ten victims, septicemic plague is virtually 100 percent deadly) they at least died fast. They weren't tortured with pain for a week or more, nor did they go insane, as thousands of citizens of Constantinople did, leaping into the sea in the hope of ending their suffering.

As a dying body becomes incapable of replacing deoxygenated with oxygenated blood, the muscles surrounding the human voice box become more acid. They go into spasms, and make a characteristic dry, shaking sound. The normal hum of life in Constantine's city had been overtaken by the sound of a thousand death rattles a day.

———

If the first line of defense for a single organism facing attack from the bacterium was the Innate immune system, society's first responders, then and now, were the city's doctors. By Justinian's day, the profession of medicine was as stratified as the Catholic Church. At the very highest rank were the empire's court physicians, themselves almost always aristocrats—*comites,* or counts. Below them were the empire's public doctors, those who were paid by their respective cities for their services, among whom the highest ranking, dating from the early fifth century, were the public physicians of Rome (commanded by law to "honestly attend the poor, rather than basely . . . serve the rich"[21]). Lowest in both status and income were the private practitioners of medicine.

Almost as important as a physician's position was his educational credential. What Berytus/Beirut was for lawyers, Alexandria was for physicians. There, a four-year course of study[22] was offered by professional teachers of medicine—the *iastrophists,* who were not merely scholars but indispensable consultants for any complicated case.[23] That the same city should be the port of embarkation for both disease and doctor has an element of irony.

The training that Constantinople's physicians secured, whether in Alexandria or elsewhere, was largely an immersion in the work of the the second-century philosopher-physician Galen, the Mediterranean world's dominant medical authority for more than a millennium. Galen's humoral theory of disease was little help in coping with any infectious disease, much less one as virulent as plague, but his disciples were not without resources in caring for the sick. During the sixth century, the greatest physicians of the known world, including Aetius of Amida and Alexander of Tralles (brother of the great architect Anthemius), resided in Justinian's empire, though not always in his capital. (Alexander eventually settled in Rome.)[24] Alexander's pharmacopoeia, and those of his contemporaries, were heavily weighted toward spells, folk remedies, and charms, including the use of gladiator's blood to treat epilepsy.[25] Cold water treatments had been a popular cure since the founding of the empire—Augustus, particularly, is said to have favored it—and one of the most famous physicians during the reign of Leo, Jacobus, earned his use-name, Psychristus, for his enthusiasm for the treatment as a sovereign remedy. Another clinical tactic much in demand was the treatment of disease by the application of material that had been blessed by a saint, preferably a hermit. These literal "blessings," or *eulogia,* were frequently no more than dust or red clay that had been touched by a holy ascetic.[26] Others included magical amulets and rings (frequently carrying the image of the biblical King Solomon).

Cold water, saint's relics, and magic amulets offered only the relief found in placebos, but—like placebos—they were also generally harmless. The same cannot be said of drugs that were known to the physicians of late antiquity, who spent much of their training in their use, though the training was not without its magical components. Galen's drug categories were based on a mild sort of sympathetic magic; the seventh-century bishop St. Isidore of Seville described the theory behind it so: "Every cure is brought about either by the use of contraries or by the use of similars. By contraries, [we mean that] a chilling disease is treated with heat or a dry one with moisture."[27] But the contraries and similars were frequently powerful alkaloids like the atropine found in mandrake and belladonna, purgatives like copper oxyacetate (verdigris) or the "juice of the opium poppy" all of which were in wide use in sixth-century Constantinople.[28] The seventh-century Alexandrian physician Paul of Aegina, in contrast, did far less damage, and probably an equal amount of good, with his belief that a homely substance like butter was useful for reducing the swelling caused by the plague's distinctive buboes.[29]

A modern reader must guard against a smugness about primitive medical practices. Physician and author Lewis Thomas, remembering his own medical education in the first half of the twentieth century (Dr. Thomas graduated from medical school in 1933, and practiced almost until his death sixty years later) writes movingly of the limited therapeutic lessons of the medicine of his own day: A doctor's job, in 1933, was to "diagnose, enlist the best possible nursing care, explain things to the patient and family, and stand by."[30] In this, he was far closer to the practice of medicine described by Alexander in his twelve-volume *Therapeutika* than to that employed by the time of the Second World War. A modern physician appalled by the use of charms and folk remedies would nonetheless find Alexander's approach to medicine familiar, particularly his respect for *peira,* the experience of a clinician willing to try anything that might alleviate the discomfort of his patients . . . even if sometimes that meant using magic charms to treat pain for a patient unwilling to take medicine orally or rectally.[31]

The limitations of the physicians of late antiquity would be shared by their successors well into the twentieth century, for until the discovery of broad-spectrum antibiotics the only weapons available to combat deadly infections were those of the body's own immune system.* One in three victims—those with a combination of good fortune, strong underlying health, and an uncompromised immune system—survived an infection of bubonic plague during those horrific months. One of them was the emperor himself.

Justinian's exposure and survival is evidence of both the egalitarian character of the disease's vector, and the vigorous nature of his body's defenses. The specific progress of Justinian's illness is undocumented, but he seems to have escaped the worst of the uncontrollable inflammatory response caused by the disease's endotoxins. As a result, though he spent weeks on the brink of death (during which time Theodora was effectively ruling the entire empire) and his demise was regularly rumored, he survived both the disease and the problematic ministrations of his physicians.

The tools available to those physicians attempting to treat the disease clinically were only slightly less effective than their ability to control its spread. Attempts to treat infectious disease by traditional public health measures, including quarantine, were in wide use by the sixth century, though problematic in effect; when Bishop Nicholas of Sion banned farmers from entering his town on market days in order to limit the spread of the disease, he was nearly arrested by the municipal authorities, who believed he was manufacturing a famine in order to drive up prices.[32] More effective, because more widely available, was the one great medical innovation of late antiquity: the hospital, which evolved from the Christian *xenodochia*, combinations of treatment centers, hospices, and poorhouses, which spread from their original locations in Judaea during the late fourth century to Rome, Ephesus, and (largest of all) the hospitals built and staffed by St. John Chrysostom during his tenure as Bishop of Constantinople.[33]

Justinian himself terminated the state subsidy paid directly to

---

* Even the most famous such discovery, penicillin, is unhelpful in combating plague, since that class of antibiotics works by blocking the enzymes—transpeptidase and others—needed for the assembly of the peptidoglycan that forms the cell wall of gram-positive bacteria only. The lipopolysaccharide-containing membrane of gram-negative bacteria, like *Y. pestis*, requires a different sort of antibiotic treatment, generally from the tetracycline family.

physicians and transferred them to the hospitals, which were by then functioning very like modern medical centers.[34] The largest of them, St. Pantalaimon, had been built on the site of the house occupied by Theodora upon her return to Constantinople and modesty—the place where she spent her last days before meeting Justinian, famously, if legendarily, spinning wool. But large as Constantinople's hospitals were, they were overwhelmed within weeks by victims of the demon, for whom they could do little except house them until they were ready for burial.

————

On the first hill of the city, at the south end of the Bosporus where Constantinople's peninsula curls like the bottom of a letter C, stands the Hippodrome, Hagia Sophia, and the emperor's palace. From that hill, Justinian and his ministers watched the destruction of their capital. A city the size of Constantinople has a normal death rate estimated at thirty a day. Increasing that by two orders of magnitude put a crushing burden on the business of easing the departed on their way to heaven. Fifty years later, Bishop Gregory of Tours would write about the pestilence,

> Since soon no coffins or biers were left, six and even more persons were buried together in the same grave. One Sunday, three hundred corpses were counted in Saint Peter's basilica [in Clermont] alone.[35]

This was nothing compared to the problem facing Justinian. In very short order, the existing burial grounds were filled, then every square foot of new ground; gigantic new cemeteries were built across the Golden Horn at Sycae. At the same time, the population that was filling up the graves rapidly overtook the population that could dig them—those who were still healthy, and not spending every waking hour caring for victims. Though burial had always been a family responsibility, Justinian could not ignore the problem, and detailed a minister named Theodorus to find a solution. A Christian city could not contemplate cremation. Instead, Theodorus looked to the walls.

Eighty years after Constantine's death, when Alaric's Goths sacked

Rome itself, the ministers of the eastern emperor, Theodosius, began construction of an immense series of walls that would surround and protect the imperial city from a similar fate. The Walls of Theodosius ran roughly north to south guarding the eastern portion of the city's peninsula from landward attack, and were nowhere less than twenty, and more frequently thirty feet high. Every 180 feet, a square tower sixty feet high was built from which Constantinople's bowmen could defend the city from any barbarian attack. The cemeteries at Sycae were likewise surrounded by such towers, and at Theodorus's direction, Justinian's troops removed the tops of dozens of the towers, and filled them with the bodies of the dead. "As a result," Procopius writes, "an evil stench pervaded the city and distressed the inhabitants still more, especially when the wind blew fresh from that quarter."

The internment of the dead was practically the only occupation that drew the city's population out into the streets, which were otherwise desolate; as a consequence, the ovens of the city's bakeries remained unlit. Justinian had constructed dozens of granaries and cisterns as insurance against another Nika Revolt, but without bakers to turn wheat and water into bread, his prudence counted for little. "Indeed, in a city which was simply abounding in all good things, starvation almost absolute was running riot. . . ."[36] In an epidemic with a direct path of contagion, the lack of human-to-human contact might have exhausted the demon in days. Human starvation, however, did nothing for the rats and fleas except provide them with a huge new source of food.

———

As with so many of the events of Justinian's reign, the most vivid chronicle of the plague's arrival is that of Procopius, whose greatest talent as a historian may have been the good fortune to be present at the most significant events of his time. This is not to diminish his other abilities, including devotion to accuracy and a clear prose style, consciously modeled on that of Thucydides (despite a pronounced taste for Homeric excess). Nor is it to grant him any sort of consistent impartiality, which is one virtue he failed to exhibit in either his sycophantic record of Justinian's architectural achievements—*On Buildings*—or his extraordinarily splenetic catalog of the moral failings of emperor, empress, and everyone else, the *Anekdota*, or *The Secret History.**

In the eight volumes entitled *History of the Wars*, however, he approaches the level of his idol, Thucydides, earning a Gibbonesque encomium: "His facts are collected from the personal experience and free conversation of a soldier, a statesman, and a traveler; his style continually aspires, and often attains to the merit of strength and eloquence."[37] It is in *History of the Wars* that his account of the plague appears, beginning with the famous lines:

> During these times, there was a pestilence, by which the whole human race came near to being annihilated. Now in the case of all other scourges sent from heaven some explanation of a cause might be given by daring men, such as the many theories propounded by those who are clever in these matters, for they love to conjure up causes which are absolutely incomprehensible to man ... but for this calamity, it is quite impossible either to express in words or to conceive in thought any explanation, except indeed to refer it to God.[38]

The description of the disease itself is a model of clinical description:

> [Victims] "had a sudden fever, [though] the body showed no change from its previous color, nor was it hot as might be expected, but of such a languid sort that neither the sick themselves nor a physician who touched them could afford any suspicion of danger.... But on the same day in some cases, the next day in others, and in the rest not many days later, a bubonic swelling developed, and this took place not only in the groin, but also inside the armpits, and in some cases also beside the ears, and at different points on the thighs.... There ensued with some a deep coma, with others a violent delirium ... who suffered from insomnia and were victims of a distorted imagination....

---

* While *On Buildings* was published during Procopius's lifetime, the *Anekdota*, for obvious reasons, remained a truly "secret" history until 1623, when a Vatican librarian named Niccolo Alemanni discovered a copy in his library.

[A]nd in those cases where neither coma nor delirium came on, the bubonic swelling became mortified and the sufferer, no longer able to endure the pain, died."[39]

Compare this with an epidemiology textbook written fifteen centuries later:

Onset is sudden with chills and rigors and rise of temperature to 102° or 103° . . . severe, splitting headache and often pain in the limbs, the back and the abdomen. [The victim] may curl away from the light, or, as the painful bubo develops, take up some attitude in bed that relieves the pressure on the painful swelling. He becomes confused, restless, irritable, or apathetic, his speech slurred as if drunken [Procopius: "the tongue did not remain unaffected . . . lisping or speaking incoherently and with difficulty"], he is unable to sleep, sometimes wild or maniacal. He may bleed into his skin, or internally into his stomach or intestine or from his kidney. . . . Within a day or two he is prostrate with all the symptoms of shock. His temperature may come down and he appears better on the third day or so, but this is deceiving; he is worse the next day and dead soon after. Most patients died between the third and sixth day.[40]

The value, to a modern reader, of such a precise eyewitness description of the demon's ravages is obvious; so obvious, in fact, that it can easily be taken for granted. But by the time that Procopius set down his chronicle, the historian's ideal of simply recording events as they happen was already starting to wane, under assault from an ever more pervasive interpretation of events through a religious, even millenarian, perspective. Those wanting to see the sixth century through a clearer lens should probably give thanks for Procopius's background and professional training, which amplified what was probably an innate and temperamental conservatism. Despite frequent temptations to record the supernatural omens and prophecies that were the stuff of daily life for many of his contemporaries, the lawyer-turned-historian was Hellenophile enough to pledge his final allegiance to the literary values of his model, Thucydides.

One of Procopius's best-known modern biographers, J. A. S. Evans, argues that the historian was a product of the same sort of selection for accent, family, and school that groomed generations of Englishmen to rule a later empire and was predictably bigoted against *arrivistes* who lacked one or more of the selected characteristics, such as Justinian, and—even more so—Theodora.

The historian's conservatism, in fact was so extreme that his writings are more respectful of the Spartans of Leonidas than the legions of Caesar, and more likely to cite the legendary heroes of Troy like Aeneas and Diomedes (who is the subject of the first story recounted upon reaching Rome in the company of Belisarius in December of 536) than Augustus, Trajan, or anyone more recent. No chronicler is a better exemplar of the intellectual passage from antiquity to medievalism taking place during the reign of Justinian.

But all accounts of the plague are revealing. Another lawyer-turned-historian from the eastern shores of the Mediterranean, Evagrius Scholasticus, was only a boy in his native Syria when the plague arrived in Constantinople, but nonetheless produced a remarkable account of the demon's spread during subsequent years, and speaks with the privileged position of one of its victims:

> [The plague] took its rise from Aethiopia, as is now reported, and made a circuit of the whole world in succession, leaving, as I suppose, no part of the human race unvisited by the disease. Some cities were so severely afflicted as to be altogether depopulated, though in other places the visitation was less violent. It neither commenced according to any fixed period, nor was the time of its cessation uniform; but it seized upon some places at the commencement of winter, others in the course of the spring, others during the summer, and in some cases, when the autumn was advanced . . . at the commencement of this calamity I was seized with what are termed buboes, while still a school-boy, and lost by its recurrence at different times several of my children, my wife, and many of my kin, as well of my domestic and country servants.[41]

Given the widespread devastation that he reports, to say nothing of the personal injury he suffered at the demon's hands, one can only call

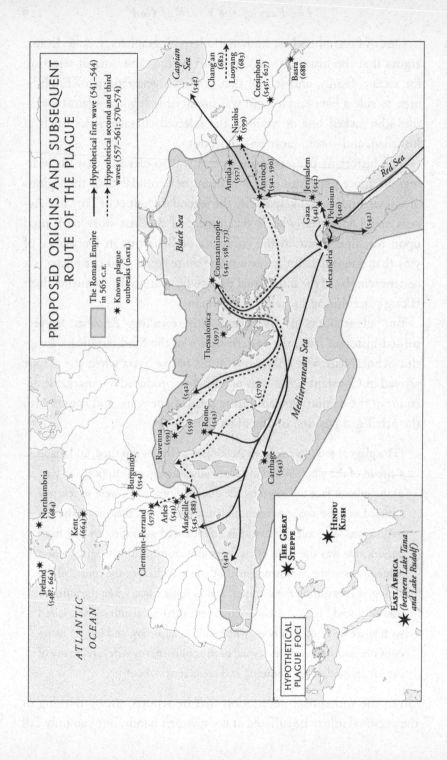

PROPOSED ORIGINS AND SUBSEQUENT
ROUTE OF THE PLAGUE

→ Hypothetical first wave (541–544)

- - → Hypothetical second and third
waves (557–561; 570–574)

▨ The Roman Empire
in 565 C.E.

✴ Known plague
outbreaks (DATE)

ATLANTIC
OCEAN

Ireland
(548?, 664)

Kent
(664)

Northumbria
(684)

Burgundy
(554)

Clermont-Ferrand
(573)

Arles
(543)

Marseille
(543, 588)

Ravenna
(591)

(559)

Rome
(543)

(570)

Carthage
(543)

Mediterranean Sea

(542)

Thessalonica
(597)

Black Sea

Constantinople
(542, 558, 573)

Amida
(557)

Nisibis
(599)

Antioch
(542, 590)

Gaza
(541)

Jerusalem
(542)

Pelusium
(540)

Alexandria

Red Sea

(542)

Caspian
Sea

(542)

Chang'an
(682)

Luoyang
(683)

Ctesiphon
(542, 627)

Basra
(688)

HYPOTHETICAL
PLAGUE FOCI

✴ THE GREAT STEPPE

✴ HINDU KUSH

✴ EAST AFRICA
(between Lake Tana
and Lake Rudolf)

Evagrius's account remarkably sedate. Though orthodox in his Christianity, his faith's taste for the apocalyptic perspective was still tempered by the restraint of its Hellenic forebears.

Not, however, everywhere. As early as the plague of 251–266, Bishop Cyprian of Carthage could cheer the disease decimating his city with a "Kill them all and let God sort them out" sermon: "How suitable, how necessary is this plague and pestilence . . . the just are called to refreshment, the unjust are carried off to torture."[42] By the time of the plague's arrival in Constantinople and the eastern Mediterranean three hundred years later, one can read:

> In this last millennium, which is the seventh, during which the kingdom of the Persians will be extirpated, the Children of Ishmael will come out of the desert of Yathrib and all come and collect there at Gab 'ot Ramta . . . While I beheld these four heads of punishments, Devastation and the Devastator, Destruction and the Destroyer, they cast lots over the land. The land of the Persians was given to Devastation for him to devastate it, sending its inhabitants to captivity, slaughter and devastation; Syria was given to the sword of Devastation, its inhabitants to captivity and to slaughter; Sicily was given to Destruction and the sword, and its inhabitants to captivity and to slaughter; the Roman empire was given to Devastation and its inhabitants to captivity and to slaughter.[43]

Perhaps the most revealing account of the plague is that of John of Ephesus, a Syrian-speaking member of a monastic order who was a professed Monophysite, and in consequence a frequent refugee from waves of persecution whenever orthodox leaders were in positions of authority in Constantinople. At other times, particularly during the ascendancy of Theodora, whose favorite he was, he enjoyed very high status indeed, even rising to a bishopric at Ephesus in Asia Minor. By his own no doubt exaggerated accounting, he was personally responsible for baptizing more than seventy thousand former pagans living on the Anatolian plateau; he has been not inaccurately described as "a writer whose zeal exceeds his elegance."[44]

His *Ecclesiastical History*, of which only fragments have survived,

contains by far the longest, and most apocalyptic, of the contemporaneous accounts of the plague. Comparing John to Procopius is an object lesson in the relative importance of prophecy over fortune:

> Thus, over these things the prophet [in this case, the author of the
> Book of Lamentations] might weep and say, "Woe upon me not be-
> cause of the destruction of the daughter of my people, but because of
> the desolation of the entire habitable earth of humanity, which has
> been corrupted by its sins; and because the world in its entirety has al-
> ready been made desolate for some time and has become empty of its
> inhabitants."[45]

and the preemption of the natural by the supernatural:

> When this plague was passing from one land to another, many people
> saw shapes of bronze boats and (figures) sitting in them resembling
> people with their heads cut off. Holding staves, also of bronze, they
> moved along on the sea and could be seen going whithersoever they
> headed. These figures were seen everywhere in a frightening fashion,
> especially at night. Like flashing bronze and like fire did they appear,
> black people without heads sitting in a glistening boat and traveling
> swiftly on the sea, so that this sight almost caused the souls of the peo-
> ple who saw it to expire.[46]

John's chronicle reads, to a modern, like an unlikely combination of
clinical accuracy and apocalyptic imagery. To his contemporaries,
however, the missionary's willingness to describe suffering on its own
terms, rather than as a way to salvation, is a pure expression of Monophysitism. To John—to many Monophysites—the argument that no
distinction could be made between the human and divine attributes
of Jesus was more than just a bit of empty theologizing. It also demanded a respect for a pragmatic understanding of disease. The demon killed and disfigured, not as evidence of God's displeasure, but
because *all* of reality is embraced in the Monophysite world.[47] Perhaps
for that reason, one of the more distinctive tropes of John's chronicle
is its strained attempt to discover a silver lining around the pestilential
cloud:

Although [the plague] was very frightening, grievous, and severe, it would not be right for us to call it not only a sign of threat and of wrath, but also a sign of grace and a call to repentance. For the scourge . . . by its silence sent as it were numerous messengers from one country to another, and from city to city and to every place, just as if somebody were to say, "Turn back and repent . . . for behold I am coming . . ."

As in the days of Noah, when that blessed man together with his family heard the message of the threat and of perdition, he grew afraid, and did not disregard it but took care to build the ark . . . So also in this time in like manner as did that blessed man, many people managed in a few days to build ships for themselves consisting of almsgiving, that these might transport them across that flood of flame.[48]

Whatever his perspective, John's eye for the telling detail is very real, indeed: Once the plague arrived in Constantinople, "nobody," he wrote, "would go out of doors without a tag upon which his name was written and which hung on his neck or his arm,"[49] a means of identification in case he died suddenly. One can, perhaps, imagine the grotesquerie of a mass grave so full that the living needed to march on top of layers of the dead in order to press the corpses more efficiently into the available space. John wrote,

How can anyone speak of or recount such a hideous sight, and who can watch this burial, even though his soul should remain in his body and not waste away from bitter lamentations over so much iniquity which would suffice to destroy the children of Adam? How and with what utterances, with what hymns, with what funeral laments and groanings should somebody mourn who has survived and witnessed this wine-press of the fury of the wrath of God?*[50]

* A direct quote from the Book of Revelation, 14:19.

PART IV

PANDEMIC

CHAPTER TEN

"A Man of Unruly Mind"

523–545

TWENTY YEARS BEFORE Constantinople's plague summer, an emissary from the Persian court arrived in the city, bearing a document intended to promote peace between two great empires, each of whose borders defined the limits of civilization. Kobad, ruler of the Persians, proposed that his son and likely successor, Khusro, be adopted by Justin, emperor of Rome.

The adoption was symbolic, of course; at the time of the proposal, in 523 C.E., Khusro was at least twenty-five years old. But adoption of just this sort was a procedure sanctified by a very long stretch of Roman history, dating back to the very first of Rome's emperors: Octavian's adoption by Julius Caesar was his entry pass into the Second Triumvirate, from which he would eventually assume supreme authority over what was still notionally the Roman Republic. Accordingly, Kobad's proposal was given due consideration by emperor, court, and Senate (including, of course, the emperor-presumptive, Justinian) and might well have been accepted, but for the argument offered by the then-Quaestor, Proclus, who pointed out that such an adoption might well give the Persian a legal claim on the Roman throne. In light of such a risk, Justin (or, probably, Justinian speaking for Justin) rejected the proposal.

This episode marked the beginning of one of history's most remarkable dual biographies, since for the next forty years the lives of Khusro and Justinian would serve as mirrors, a kind of infinite regress where every action taken by one would be reflected in a slightly distorted version taken by the other. For once, the cliché "inextricably linked" is apt.

Events would likely have linked the two under any circumstances; they were, after all, rulers of two great empires with a shared border during a time when both, as Kobad had accurately feared, faced threats from many of the same barbarian sources. But circumstances can only go so far to explain the doppelganger-like phenomenon that has occurred to virtually everyone who has ever written about the two—many of whom feel obliged to take sides: J. B. Bury, as a case in point, sniffs, "In the long competition between Khosrow* and Justinian, the advantage, both of merit and fortune, is almost always on the side of the barbarian."[1]

Bury's ranking aside, his use of the word "barbarian" was inapt not only when he wrote it in 1923, but would have been so during the sixth century as well. He was, however, only the latest in a long line of historians, beginning with Herodotus, that used that term to describe the rulers—the "Asiatic" rulers—of the lands to the east of the Mediterranean. No great physical boundary separates western Mesopotamia from the most easterly of the pre-European cultures of Greece and Rome, but geography nonetheless gives some shape to the thousand-year-long conflict between them. Seen from the Greek or Roman perspective, the west-to-east topography that begins with the Mediterranean shore of modern Lebanon and Israel immediately climbs to a north-south chain of mountains including the Amanus, the Lebanon, and the Judaean hills. Given the obstacle (some of the peaks are more than 7,500 feet/2,500 meters high) posed, within fifty miles of the Mediterranean littoral, by these mountains, the river valleys that flow through them take on considerable significance. The most important of these rivers are the two northernmost, the Orontes and the Litani, that pass through the mountains to the Mediterranean (a third river, the Jordan, flows south to the Dead Sea). Just to the east of the mountain ranges, the plateau of Jordan and Syria gradually drops from an average altitude of fifteen hundred feet to only a bit above sea level in Iraq.[2]

---

* Historians writing in western languages use half a dozen variant spellings for Khusro's name, including Cyrus, Chosroes, Xusro, and Khosro.

The result is this: What begins in the west as an extremely fertile region, particularly around the northern river valleys, rapidly becomes a desert. Cartographers mark maps with lines that connect geographic areas with the same characteristics; the most familiar are probably the contour maps that indicate altitude. Others are the ones used by meteorologists to display lines of constant air pressure (isobars) or constant temperature (isotherms). The relevant line demarking Mesopotamia is the *isohyet* . . . a line of constant precipitation. Past the Orontes and Jordan river valleys is the 200 mm isohyet line, the point where annual rainfall drops below 200 mm—a bit less than eight inches. Eight inches represents the limit below which farming without irrigation becomes impossible, with all the consequences that entails. In the same way that the presence of water defined the internal boundaries of the Mediterranean empires that grew with seaborne trade and colonization, the absence of water defined the boundaries of the Mesopotamian empires, as far back as the birth of civilization itself. The "fertile crescent" of high school geography is outlined, on its concave side, by the 200 mm isohyet, with deserts to the north and east of the valleys of the Tigris and Euphrates.[3]

One could easily make the case that imperialism was invented within the 200 mm isohyet, from the time that Sargon the Great founded his multiethnic Akkadian empire around 2200 B.C.E. and steadily refined there by Hammurabi's Babylon and Sennacherib's Assyria. Even the Achaemenid empire of Xerxes, Darius, and Cyrus, whose home was in the Iranian highlands, established their administrative capital in Susa, within the lush lowlands of Mesopotamia. For two millennia, relations between the rulers of the fertile crescent and their Mediterranean neighbors cycled between disregard and wars of conquest, thus giving the cultures of the west their first opportunity to define themselves against a hostile enemy. Only during the century following Alexander's defeat of Darius at Gaugamela in 331 B.C.E. were the rulers of Mesopotamia culturally affiliated with the Greek speakers that dominated the Mediterranean, since the empire founded after Alexander's death by the Macedonian general Seleucus Nicator was, unsurprisingly, Hellenophilic. Even the Parthians who gradually supplanted the Seleucids a century later were admirers of Mediterranean culture, despite some significant wars with the new and vital Mediterranean empire of Rome.

So things stood until the third century, when Mesopotamia, like Rome, faced an empire-shaking crisis. And, like Rome, it ended up stronger but changed.

In the year 224, a decade before the assassination of Severus Alexander inaugurated Rome's Third-Century Crisis, a minor noble from the Fars family of the Iranian highlands named Ardashir Papakin revolted against his Parthian overlords and, rather quickly, replaced them with a dynasty called the Sassanids. To legitimize his rule, the Iranian claimed descent from the most notable ruling dynasty the region had ever known, the Achaemenids of Cyrus the Great. The names Papakin and Sassanid are clues to the fabricated nature of his borrowed glory. Subsequent state-approved histories chronicle how an Achaemenid king named Papak dreamed that the sons of his shepherd, Sasan, would rule the world; so inspired, the king gave his daughter (in some versions, including Agathias's sixth-century history, his wife) to Sasan. More favored by modern historians, though lacking some romance, is the likelihood that Sasan was a minor Fars noble, Papak his son, and Ardashir his grandson. Whatever the story, the family bequeathed to subsequent generations the names of a language—Farsi—and a nation—Persia. They also left them a newly aggressive attitude toward their imperial neighbor to the west.

Shapur I, who defeated and captured the emperor Valerian in 260, described himself as "king of kings of Iran and non-Iran, of the race of Gods . . . king of kings of the Iranians, of the race of gods, grandson of Papak, king; of the empire of Iran, sovereign."[4] Ammianus Marcellinus quotes a letter in which Shapur II is self-described as "king of kings, partner of the stars, brother of the sun and moon."[5] The progression from minor nobility, to regional royalty, to the pages-long titles claimed by Shapur, marks increasing confidence for the Sassanid Persian empire, which extended its western border into Armenia— eventually divided, like eighteenth-century Poland, between Roman- and Persian-dominated halves—and its eastern boundaries into mod-

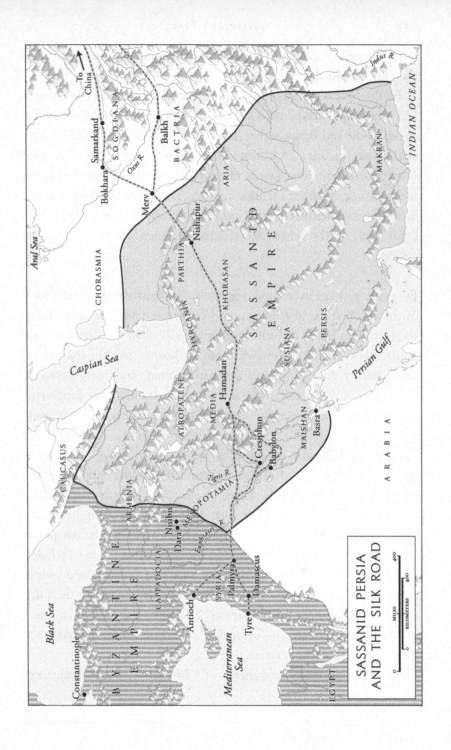

SASSANID PERSIA
AND THE SILK ROAD

ern Afghanistan and across the Khyber Pass into Pakistan, or, as it was then known, Kush.

By the fourth century, the two great empires were becoming more and more alike. It is perhaps too easy to find points of comparison between Rome and Persia in their last centuries, and one should be skeptical about theories of history that depend on like equaling like. Nonetheless, one cannot write of either empire during the fourth and fifth centuries without mentioning the strains they faced, on the one hand, from sometimes conflict-ridden marriages between state and religion and, on the other, by mounted bowmen out of the Eurasian steppe.

————

By some, perhaps overstated, measures, "Zoroastrianism is the oldest of the revealed world-religions, and it has probably had more influence on mankind, directly and indirectly, than any other single faith."[6] The religion's somewhat murky history reaches back to its sixth-century B.C.E. founder, Zoroaster (often, and more correctly, Zarathustra), whose doctrines, including salvation, a "last-judgment" resurrection, and eternal life for the faithful,[7] anticipate Christianity by six centuries. The fiercely dualistic religion—Zoroastrianism theodicy explains the existence of evil by postulating a supremely good deity, Ahura Mazda, in eternal conflict with the dark god, Ahriman—had been virtually exiled from Zarathustra's Azerbaijani/Iranian home for five hundred years after Alexander's conquest,* and might well have remained so. The first Sassanids had taken power without an orthodox religion, and to the degree they shared a religion at all it was likely to be the syncretic one founded by the Mesopotamian Mani, which drew from Gnosticism, Mithraism, Christianity, and even Buddhism. When Bahram I came to the Sassanid throne, however, he exiled Mani, banned Manichaeanism, and established Zoroastrianism as the state religion.[8]

As with Constantine's embrace of Christianity, the consequences

————

* The religion's place of exile, northern India, remains the largest Zoroastrian community in the modern world, where its practitioners are called Parsis . . . Persians.

were large. A system of relatively immobile birth castes—priests, soldiers, scholars, artisans—became state policy. The "Letter of Tansar," popularly supposed to have been written by Bahram's chief priest,[9] ordered "there shall be no passing from one [birth caste] to another unless in the character of one of us outstanding capacity is found."[10] Construction of some of the most distinctive architectural forms of the Sassanid period, the "fire shrines," which equated fire with light with good with Ahura Mazda, became a state imperative. "Each king apparently had his own fire, lighted at the beginning of his reign, and this fire was on a portable fire altar."[11] And, in Persia no less than Rome, a hierarchy of priests, or *magi,* like Christian bishops responsible for religious administration in geographic districts, accumulated significant potential political power.

That power might have remained potential rather than actual but for the entry of another set of actors onto the Persian stage, ones that would have been as familiar to Rome as doctrinal conflicts between powerful clerics. If the Sassanids were to have their own Constantine, they must also have their own Attila.

The origins of the Ephthalite, or "White" Huns who arrived on the borders of Persia's empire in the middle of the fifth century are no better documented than the birthplace of Attila's predecessors. The name itself isn't much help; just as every Central Asian steppe dweller from the second century B.C.E. appears in contemporaneous chronicles as a Scythian, so every one from the fourth century C.E. is a Hun (and every one from the tenth to the fourteenth century is either a Turk, a Tartar, or a Mongol). The Ephthalites *might,* like the Huns, have originated in the Altai mountains and spoken a language from the Altaic family, but they equally well might not. Either way, they conquered the Kushan tributary nation of Persia sometime in the late fourth century, and by the fifth century represented a clear danger to the Persian state.

The danger peaked during the reign of Firuz, the *shahanshah,* or king of kings, who took the throne in 457. Firuz was humiliated at the hands of the White Huns when he was captured at the end of a failed punitive expedition—Firuz's advisers convinced him that he could bow before the rising sun, symbol of Ahura Mazda, letting the Huns think he was prostrating himself before them. The scheme didn't work, and Firuz was ransomed and sent home, chastened.

The chastisement did not, evidently, take, since the Persian king returned to battle against the White Huns in 484, only to be defeated again, this time by a clever ruse: The Huns dug a trench in front of their defensive lines, covering it with reeds and dirt, so that the entire Persian host charged into the trench only to be massacred. Firuz was not merely defeated, but killed. Once he knew all was lost, he tore the great pearl from his ear in order to prevent any White Hun from wearing it as the spoils of battle.[12]

Kobad, Firuz's youngest son (Procopius mentions thirty) also became his heir, but the throne he inherited was shaky indeed. Not only was the young prince facing a formidable invasion from the steppe but also an increasingly restive aristocracy and an ever more powerful episcopate. As a counterweight, the new ruler turned to a group of populist religious reformers. The doctrines of the Mazdakites, so called for their founder, a Zoroastrian priest named Mazdak, included redistribution of wealth, free love—Procopius called it communal intercourse—and, most incendiary of all, a loss of power for both magi and nobles.

Unsurprisingly, the nobles rebelled against their new king, jailing Kobad. His notably ignominious escape—the king disguised himself as his wife, whom he had persuaded to gain access to her husband's prison by offering sexual favors to his jailers—led him to a new ally: the White Huns themselves, available for a price. Apparently, their services could be purchased on credit, but when the time came to redeem his promissory note, Kobad was unable to do so and attempted to raise the money by borrowing it from his opposite number in Constantinople, the Emperor Anastasius. Rebuffed by the emperor, Kobad, "saying that [the Romans] had caused the incursion of the Huns, and the pillage and the devastation of their country,"[13] raided imperial territory in 502 with the object of stealing that which he could not borrow. By the time the Persian king had stabilized his relationship with the Huns, and proposed imperial adoption for the Persian crown prince as a peace gesture toward Anastasius's successor, Justin, the once-quiet Mesopotamian border was becoming an increasingly precarious place.

———

Fifteen centuries after the dramas of late antiquity played themselves out, we strain to translate the now-exotic concepts that defined Justinian's world. Even so, we have a good deal of cultural continuity with that world. The same cannot be said for the empire of the last great Persian ruler, Khusro the Great; unbeknownst to the Sassanids, one of the great discontinuities in human history, the complete destruction of the Persian empire by the armies of Islam, was looming less than a century hence, as was the retrospective rewriting of history that is the privilege of conquerors. One of the reasons that Khusro looms so large in all subsequent chronicles is that virtually all the architectural advances, military triumphs, or cultural achievements of pre-Islamic Iran are often attributed to him: It was easier to fasten the glories of Persia's imperial past to a single standard.[14]

But it is not the only reason for Khusro's reputation. Two of the most vivid representations believed to be of Khusro appear on a silver plate in the Hermitage Museum, in which the king appears in not only a formal and flat display of the ruler surrounded by his dignitaries (not so very different from the mosaic of Justinian at San Vitale) but also astride a horse, turned backward, bow in hand, to make what is traditionally called the Parthian Shot at two rams,[15] a telling reminder that Persia's last great ruler carried the weight of all his predecessors.

By the time Khusro was in his early twenties—his birthdate is unknown but believed to be around 496—he was not only the heir apparent, despite several older brothers with equally legitimate claims on the Persian throne, but a power in his own right. In fact, among the literally dozens of points of resemblance with Justinian is Khusro's dominance of national politics before his ascendance to the throne. In 528, still the crown prince, not yet king, Khusro discovered that his father's Mazdakite allies were conspiring against the throne. Driven, perhaps, by a combination of loyalty, anger, and a desire to demonstrate a kingly sort of resolution, in 529 the prince arrested, tortured, and executed Mazdak, and followed up with a massacre of his followers. (The Mazdakites would one day serve as inspiration for Islam's dissident Shi'a.)[16]

By then, Khusro's lifelong obsession with Justinian was well developed, and generations of subsequent readers have been struck by what appears to be his eagerness to return the slight visited by Justin's court on Kobad's plan for adoption. Whether overblown or not—because so many of the Persian accounts of Khusro's reign were either destroyed or rewritten by the Sassanids' Arab conquerors, a disproportionate amount of information comes from the work of western chroniclers like Procopius and Agathias—the first seeming opportunity for revenge came almost on the heels of the Mazdakite massacre, and from an unlikely place.

The Academy of Athens, founded by Plato himself, had been the world's greatest center for philosophical inquiry for a thousand years when Justinian took the throne. It was not only prestigious, but both very rich and very pagan—all provocations to an emperor who was avid for legitimacy, orthodox religion, and gold. In 529, as part of an imperial ban against pagan education,[17] the Academy was shut down, and while the members of its faculty were offered pensions and resettlement, seven of them—Damascius, the Academy's head; Simplicius; Eulamius; Priscian; Hermeias; Diogenes; and Isidore—were recruited by Khusro to re-create the Academy at the Sassanid capital city of Ctesiphon, there to translate the works of Plato and his successors into Persian.[18]

The scholars, in retrospect, seem a bit naïve; raised on Plato, they were perhaps susceptible to an invitation from what they believed to be, in the words of Agathias, "the land of 'Plato's philosopher-king' in which justice reigned supreme."[19] However, Khusro's claim on the title is considerably stronger than most. In the *Karnamag*, an Arabic redaction of a book originally written as *The Book of the Deeds of Khusro*, the king is quoted saying, "The most disgraceful thing for kings is to disdain learning and be afraid of science."[20] One of the few surviving works created by the transplanted scholars, Priscian's *Solutiones ad Chosroem*, is a classic pagan Neoplatonist text, touching on natural history, ontology, and the mind-body relationship, structured as a philosophical debate at Khusro's court, which flatters the king by making him one of the interlocutors—a flattery that makes no sense for a king who did *not* style himself a philosopher.

Priscian's work, though a minor one, is evidence that, despite the significant barriers presented by state-enforced orthodoxies in both the Persian and Roman empires—significantly, far better enforced in the latter than the former—the basic tools of western philosophical inquiry continued to find their way east. Though the home offered by the philosopher-king to his expatriate Platonists proved temporary—by 532, homesick, all had returned to Athens*—it is a reminder that the seeds planted in Athens and nurtured in Mesopotamia were, before being rediscovered by medieval Europe, diffused along some underappreciated paths. One of them led through the Sassanid court.

When Kobad died in September of 531, Khusro, then thirty-five years old, had had a decade to prepare for the moment, and successfully maneuvered around the court's nobles and the claims of his older brother to take his place as "sovereign of Iran and the entire corporeal world . . . divinely designated protector, and guide of the material creation."[21]

In 532, in return for 11,000 pounds of gold, putatively for taking responsibility for defending the Caucasus against barbarian invasions, Khusro signed the so-called treaty of Endless Peace with Justinian, who was newly confident after surviving the Nika riots and eager to send his armies west without having to worry about his eastern flank. Because so many of the surviving accounts were written from the Roman perspective, information about Khusro and his empire during the seven years after execution of the treaty is sparse. A reasonable guess, however, is that Persia's new ruler occupied himself largely with reforming Persia: its administration, its army, its treasury, and its capital.

To Khusro, such categories were not discrete. One of the relatively

---

* According to Agathias, the most extraordinary and disgusting aspect of Persian life was "that even though a man could and did have any number of wives, people still had the effrontery to commit adultery."

few maxims that can be reliably attributed to him reads, "The throne depends on the army, the army on revenue, revenue on agriculture, and agriculture on justice."[22] When Khusro took the throne, the standing army he inherited, some seventy thousand cavalry plus infantry and other retainers,[23] reported to a single generalissimo. The new *shahanshah* reorganized the vast territories of his domain into four districts, each with an army commanded by a general-in-chief, or *spahbedh*, and a civil administration by a governor, or *padehospan*, with the latter subservient to the former. Despite their hierarchical disadvantage, the local governors (or, rather, their improved efficiency as tax collectors) seem to have been the object of Khusro's reorganization. As with Justinian's early selection of, and lasting loyalty to, John the Cappadocian, Khusro recognized that his imperial ambitions would have to be delayed until they could be afforded. An Arab chronicler writing five centuries after Khusro quotes him saying, "I . . . assembled the officials and the taxpayers, and from their confusion I realized that there was no remedy for the situation except making the taxes just and fixing their rates on every town, on every district . . . and on every man."[24]

With an assured revenue stream that was not only efficient but generally regarded as legitimate by its payers, Khusro was able to turn his attention to his army. His military reforms proved not only more profound but more durable. Under previous sovereigns, Persia's army resembled a feudal levy of medieval Europe, with landed nobles providing and arming their own soldiers, from archers to field officers. Once Khusro was able to pay his commanders directly, he wasted little time in doing so. His "knights," or *dekkans*, were paid not in gold, however, but in land; and whether by accident or design, the ascendance of the *dekkan* created a new class of landowners (though of relatively small holdings, usually villages that Khusro granted them in fief),[25] who not merely owned land and provided military service to the king but also became a class of bureaucrats that would, in time, administer the empire.

Long before that, however, their common recruitment, training, and equipment turned them into a far more disciplined and, therefore, effective military. By the sixth century, the professionalization of the Persian army was complete. At its core, as with Rome under Justinian, were heavy cavalrymen—the Immortals, a brigade of which Belisarius had defeated at Dara—who were uniformly armed and armored with "helmet, hauberk (in Pahlavi [or Middle Persian] *griwban*), breastplate, mail, gauntlet (*abdast*), girdle, thighguards (*ran-ban*) lance, sword, battle-axe, mace, bowcase with two bows and two bowstrings, quiver with 30 arrows, two extra bowstrings, spear, horse-armor . . . and a lasso (*karmand*) or a sling with slingstones."[26]

As would shortly become evident to Justinian's generals, the army was the instrument of Khusro's ambition, not the object. Though training, equipping, and paying for his army consumed much of the *shahanshah*'s energy, he was still able to find time to place his mark on every aspect of Persia's national life, from religion to astronomy. The tower of Khusro's achievements rose high enough, in fact, to be visible even over the historical wall built by his Islamic successors. His *dekkan* system was not merely preserved by Persia's Arab conquerors but would eventually appear as the model for European feudal vassalage (more because of convergence than shared ancestry). Similarly, while the sacred book of Zoroastrianism, the *Avesta,* was codified under Khusro's reign, it was his sponsorship of the definitive edition of Persia's national epic, the *Khwataynamak* ("Book of Kings"), that better survived the Arab conquest; hundreds of years after Khusro's death, it was reproduced by Ferdowsi, Iran's greatest epic poet, under the title *Shah-Nameh.*[27] Khusro was likewise a patron of astronomy, and the star tables produced during his reign, the *zaij-I-Shahriyar,* remained in use for centuries. Similarly, the great medical school that he built at Jondeshapur was "the model on which all later Islamic Medical Schools and hospitals were to be built."*[28]

Just as Justinian is inconceivable without Constantinople, so Khusro was both created by, and creator of, his royal city—Kisfun, or, as it

---

* One of the reasons for Jondeshapur's prominence was that Khusro, believing that the medical state of the art was to be found to the east of his Kushite satellite kingdom, sent his court physician to India to collect Sanskrit medical texts. In legend, at least, he returned not merely with a medical library, but a sackful of figurines and a square board made up of sixty-four squares, half dark and half light. The game of chess remains one of the *shahanshah*'s least politically important, but most enduring, bequests to the medieval world.

appears in its Romanized form, Ctesiphon. Ctesiphon had been built on what was originally the site of an Assyrian city, Opis, which stood next to the Tigris River and astride the "Royal Road" connecting it to the Assyrian capital of Susa. Inherited successively by Babylonian and Achaemenid Persian rulers, it then passed into the hands of the dynasty founded by Alexander's general Seleucus upon the Macedonian conqueror's death. He, in turn, built a new capital, Seleucia, on the other side of the river from Opis, from which yet another brief empire was ruled. When the Parthians displaced the Seleucids, Opis was the site for their western capital, Ctesiphon, and after the Sassanids conquered the Parthians, they turned the Parthian regional center into their capital, preferring the rich provinces of Mesopotamia to their historical but barren home in the Iranian highlands.

The city's proximity to the eastern borders of Rome's empire made Ctesiphon an almost irresistibly tempting target, and the city was captured three times during the second century C.E. in the course of campaigns by Trajan, Lucius Verus, and Septimius Severus—the Arch of Septimius Severus in Rome memorializes his capture of Ctesiphon in 198—and again in 283 by the emperor Carus. Julian the Apostate died in his retreat from a failed attempt to capture it.

By then, the city had been the first city in the various Persian, Parthian, and Sassanid empires for a millennium. And it must have been spectacular. All that remains from the great palace built by Khusro is the *Taq-e-kisra*, or Great Arch of Ctesiphon, 115 feet high and 82 feet wide, and 150 feet deep, which until the construction of the Gateway Arch in St. Louis was the largest unsupported parabolic arch in the world, and is still the widest span of unsupported brick. The Taq had served as entry into the palace's great hall or *ayvan*, one of two such,* each roofed with a barrel-vault. The floor was covered by a huge silk carpet, nearly three hundred feet long, illustrating a garden, and embroidered with gold and pearls. As was traditional, the carpet had a name: "Spring of Khusro," in honor of its subject matter; in 1977, the composer Morton Feldman wrote a sonata for piano and

---

* A nineteenth-century engraving shows the full façade, with both halls, before one was destroyed in an 1888 flood.

violin entitled "Spring of Chosroes" in homage to the legendary carpet. After the sack of Ctesiphon in the late seventh century, Arab historians described the carpet in detail, as well as the gold-bound mosaics on the palace's upper walls, and the marble in all the colors of the rainbow that lined the lower walls.

The decoration for the palace was the highest art known to the Sassanids, whose most distinctive and exemplary form was architecture—and monumental architecture, at that.[29] But unlike the great buildings of Justinian's capital, most especially the Hagia Sophia, the structural logic of Sassanian architecture was less aesthetically important than its decoration, frequently images of flowers or simple geometric forms.[30] The stylized geometry of Islamic art, everywhere from Cordoba to Kandahar, still carries the imprint of the floral and geometric styles of the Sassanids who ruled pre-Islamic Persia.

Ctesiphon, like Constantinople, was simultaneously an intellectual, literary, religious, and commercial center, overseeing a huge expansion of trade during late antiquity, especially the high-margin international trade in luxury goods, particularly silk and objets d'art. Khusro's empire "dominated international trade, both in the Indian Ocean and in Central Asia and South Russia,"[31] and the best-known Sassanid motif, the mythical phoenix-like bird known as the *senmurv*, appears on art objects as far from the Iranian plateau as India and China.[32]

Philosopher-king, builder, warrior prince, patron to physicians and astronomers, Khusro enjoyed a reputation—he well earned his surname *Anushirvan*, or "Immortal Soul"—that would have loomed large even had he not lived on the other side of Justinian's mirror. But while it seems certain that the two rulers were never far out of each other's thoughts, it was not until 539 that their lives again intersected. In that year, two men, dressed in the vestments of Arian priests, arrived at Khusro's court at Ctesiphon. They were, in fact, agents of the Ostrogoth's king, Vitigis, who importuned the Persian ruler to open a second front against Justinian, and whose own troops were at that moment marching up the Po River valley. Khusro received them—according to Procopius, he solicited them—at the same time he was meeting with representatives from Armenia, who presented him with a similar plea. The Armenians were historically a buffer state and consequently nonaligned as a matter of survival. They had, however, been pushed into the Persian camp by Justinian's response to an Armenian insurrection, in which the murderers of one of Rome's supporters had escaped into Persia. And if that were not enough to tempt Khusro to action, he had reason to believe that Justinian was attempting to suborn an Arab ally of Persia, the Lakhmids.

Since the third century, Sassanid rulers had supported an Arab tribe occupying strategic areas of the Syrian desert as far south as what is today northern Saudi Arabia. Their rulers—the *banu Lakhm*, westernized into Lakhmids—had been employed both to deter raids from other Arabs of the region, and as a check against the Great Power machinations of the Roman Empire. Rome, in turn, retained its own Arab clients, the Salihids, fanatically loyal Christians who had been fighting the Persians on behalf of the empire since the time of Thedosius, for precisely the same purpose.[33]

In 529, however, Justinian replaced the Salihids with another Christian Arab tribe, the Azd, and their ruling dynasty, the Ghassanid, appointing the commander of Belisarius's Arab auxiliaries at the Battle of Callinicum, al-Harith ibn Jabala, as phylarch, or imperial representative. For the next fifteen years, the Lakhmids and Ghassanids would fight a vicious proxy war driven as much by the temperaments of their respective leaders as by the policies of the great powers for whom they were agents. The Ghassanids' leader, al-Harith, in the words of a contemporary chronicler, generated "awe and terror [in] all the nomad tribes of Syria,"[34] while his opposite number, al-Mundhir ibn al-Naaman, was, according to Procopius, a man "most discreet and well experienced in matters of warfare, thoroughly faithful to the Persians, and unusually energetic—a man who, for a space of fifty years, forced the Roman state to bend the knee."[35]

So when Justinian sent two emissaries to adjudicate a dispute between al-Harith and al-Mundhir, Khusro saw it not merely as a provocation, but a casus belli. In calculating whether to break the rather unwisely named Endless Peace, the Persian ruler had to measure actual benefits against potential costs: the gratitude of the leaders of buffer states who wanted to play one great power off against another; and the likelihood of defeat at the hands of the Romans. The latter seemed at an all-time low, since most of Justinian's best generals were occupied in Italy and North Africa—Belisarius was, at this moment, on the march to Ravenna—and the most formible soldier in the east, Sittas, had been killed in a skirmish in Armenia. In any event, the risks seemed manageable, and Khusro began a war against Justinian that lasted until 545.

The first act of this latest version of the long-running Roman-Persian war took place when Khusro led his army along the left bank of the Euphrates into Syria, with the objective of taking the city of Sura in the northern Syrian desert. They succeeded not by storm but by subterfuge; after charming the city's bishop, Persian troops escorted him back to his city where the defenders, misled by the bishop's cheerful report, left the gates open long enough for the Persians to place a stone or log on the ground that prevented the gate from closing, and permitted access to the city by the Persian cavalry. As if to support the notion that the war was at least partly driven by personal malice, Khusro's first act upon capturing his first imperial city was to send Justinian's governor directly to Constantinople, specifically instructing him to tell the emperor that the Persian ruler was loose in Syria.

Having honed his appetite on the relatively modest spoils of Susa, Khusro then turned west toward the far richer prize of Antioch.

The wealthiest city of the Seleucid empire, Antioch retained its prominence after conquest by Rome in the first century B.C.E., and occupies a large space in the history of Christianity as well, as a hotbed of Arianism in the fourth century and Monophysitism in the sixth, but even more as home to biblical literalists like Theodore of Mopsuestia and especially the city's—and, later, Constantinople's—bishop, St. John Chrysostom, each of whom served as an intellectual counterweight to the even older Christian theology based in Alexandria. By virtue of its location at the mouth of the Orontes, with access to both the Mediterranean and Mesopotamia, Antioch, the most easterly of the empire's great cities, had become a trading and manufacturing center, using its positional advantage to become enormously wealthy in part by reweaving heavy Chinese silk, purchased from Persian middlemen, into the gauzy substance so beloved by Mediterranean elites.

The same easterly position that brought the city its riches made it vulnerable as well, and it had attracted Persian attention before Khusro; in the mid-third century, Ardashir's son Shapur was the first Sassanid to conquer the city. As a result, when Justinian received word of Khusro's invasion, he sent a detachment of soldiers to Antioch to evaluate the city's prospects. The mission was commanded by the emporer's cousin, Germanus.

In Procopius's words, Germanus was "a man endowed with the finest qualities and remarkable for his activity; for in war, on the one hand, he was not only a most able general but was also resourceful and independent in action, while in peace and prosperity, on the other hand, he well understood how to uphold with all firmness both the laws and the institutions of the state. As a judge he was conspicuously upright . . . a man of very impressive personality and exceedingly serious demeanor . . . [and a] pleasant and charming host."[36] Even in the *Secret History*, Procopius has nothing but praise for Germanus, possibly out of some commutative property of biliousness: Procopius despised Theodora, and Theodora loathed Germanus. Procopius's distaste for Theodora was at least partly explained by Justinian's decision to make an empress out of a woman with distinctly dubious origins, while those same origins contributed to a very strained set of in-law tensions: Germanus had married Passara, a member of the Anicii, one of the most aristocratic families in the empire.[37]

Despite Theodora's antipathy, Germanus retained Justinian's trust. He reported back to Constantinople that, in his professional military opinion, geography was against the defenders: A great rock, almost the height of the walls, loomed large on one side and was not enclosed by the city's defenses. A sufficiently large besieging army could use this rock as a combination of siege tower and scaling ladder, and were thus provided with an irresistible path into the heart of the city. The conclusion was hard to avoid: Since the Persian objectives were apparently loot rather than conquest, better to pay for the defense of the city in gold than in blood. Rejecting this logic, the bishop and prefect of Antioch decided to resist Khusro, and asked Germanus to stay and help in the city's defense. His advice scorned, correct in his belief that his small unit would be unable to help in the defense of Antioch, and evidently unconcerned about accusations of cowardice, Germanus and his soldiers escaped the city just before Khusro's arrival.

He cannot have felt much satisfaction at seeing his prediction borne out. Khusro demanded one thousand pounds of gold to leave the city alone; Antioch rejected him, resisted, and was quickly defeated and looted, with tens of thousands of its citizens either killed or enslaved. Procopius writes:

> I cannot understand why it is the will of God to exalt on high the fortunes of a man or a place, and then cast them down for no cause that we can see . . . it is not right to say that with Him all things are not done according to reason, though He then endured seeing Antioch brought to the ground at the hands of a most unholy man.[38]

Growing more confident with each easy victory, Khusro then turned to the northeast, on a march whose objectives were the extortion of money from each walled city en route. With the example of Antioch preceding him, the need for pitched battles was nonexistent. Edessa was good for two hundred pounds of gold, Dara for one thousand pounds of silver. In 541, Khusro returned to his home in Mesopotamia, quite properly feeling he had earned the right to boast. The form of that boast was as distinctive as the man himself: He built an entirely new city near Ctesiphon, settled the Antiochene captives in a simulacrum of their home city, complete with hippodrome and baths, and named it the Antioch of Khusro (more specifically, and arrogantly, Khusro's Better-than-Antioch).

By 542, the Persian's appetite for Roman gold had returned, and once again he led an invading army across the Euphrates. But this time, he faced two obstacles that had been absent during his last excursion. One was Belisarius, recalled by Justinian from his conquests in Africa and Italy, and sent east to confront Persia again. In

Procopius's account, the Thracian general duped Khusro into retiring back across the Euphrates by dressing some six thousand of his biggest soldiers in civilian dress in order to show their complete disregard for the Persian army. In the words of Belisarius's greatest twentieth-century admirer, Basil Liddell Hart, the general, who had already demonstrated "how to provoke the barbarian armies of the west into indulging their natural instinct for direct assault," applied a different approach to the Persians: "He was able at first to take advantage of their feeling of superiority to the Byzantines, and later, when they learnt respect for him, he exploited their wariness as a means of out-maneuvering them."[39] But even without Justinian's greatest general standing in Khusro's path, his expedition was problematic. As J. B. Bury observed eighty years ago, "there existed a sufficient cause, unconnected with the Romans, to induce his return to Persia, namely the outbreak of the Plague."[40]

———————

Modern scholars have traced the path of the plague east by careful readings of everything from gravesites to notarized wills. After Pelusium, as reported by John of Ephesus, the disease next arrived in the Mediterranean coastal city of Gaza in the last quarter of 541. A dozen separate funerary inscriptions in the Negev cities of Nessana (due south of Gaza along the route of one of the wadi, or shallow streambeds, that crisscross the deserts of the Middle East) and Beersheba trace the southward journey of rats, fleas, and bacteria throughout 541; while, simultaneously, the plague spread north, to the coastal cities of Ashkelon, Ashdod, and Rehovot.[41] In early 542, a contemporary account describes the retreat of a Christian hermit to the Judean hills outside Jerusalem, an escape from the "great and most terrible mortality"[42] that afflicted the holy city. Slightly later in 542, the city of Antioch, already looted and depopulated by Khusro's soldiers, now greeted an army of *Y. pestis*: "In the whole region and in the city of Antioch, people were smitten with disease in the groin and armpits and were dying . . ."[43]

Despite such careful research, the distortions that have accumulated in the historical record over the fifteen intervening centuries have fueled debate over the path of the plague, its impact, and even its source. While hardly any historian ignores the fact of the pandemic, some have questioned whether the guilty pathogen is the same as the one that causes bubonic plague.

The reasons for the residual controversy about the etiology of the pandemic are somewhat circular. Because humanity's three great plague pandemics—in addition to the Justinianic Plague, they are the Black Death of the fourteenth century, and the late nineteenth-century Chinese plague—are widely believed to share the same source, anything that casts doubt on any episode compromises the others. In the first such investigation, in the early 1980s, the historian Graham Twigg used the then-current belief that black rats were absent from many parts of northern Europe during the fourteenth century to argue that the Black Death was anthrax, rather than plague.[44] When his arguments were expanded to encompass the sixth-century pandemic as well, it was rather as if a history of the causes of World War II was used to reevaluate the causes of World War I. Other studies drew upon English parish death records to question both the Black Death's duration of infectiveness, and the probability of transfer from an infective source to a susceptible host. Since two of the variables used to calculate the disease's basic reproductive number were inconsistent with what was known about the plague during its nineteenth-century reappearance, the authors of these studies argued, once again retrospectively, that since modern plague behaved differently from medieval plague, then the plague of late antiquity must be viewed skeptically as well.[45] They were helped by a seemingly too-rapid spread of the disease by its very sedentary rodent hosts, and by a relatively small number of discoveries of *Y. pestis* DNA in the graves of victims.

The revisionist arguments about the disease, however, while clever and well reasoned, have persuaded few historians, and provoked many, and the provocation has marshaled even more telling arguments in favor of *Y. pestis,* relying on both the bacterium's molecular history[46] and especially its ecology.[47]

That one should need such arguments at all seems slightly odd, given the extraordinarily precise description of the disease's most characteristic sign—the inguinal buboes—by historians like Eva-grius and Procopius, who were close enough to touch them. That fall, the buboes made their appearance in the Persian army, and, in com-bination with Belisarius's generalship, persuaded Khusro to retreat back across the Euphrates into his own lands, thus far free of the dis-ease. A year later, in 543, Khusro invaded Roman Armenia—today's Azerbaijan—but did not meet with the ambassadors sent by Justinian to sue for peace because one caught the plague en route. The Persian ruler immediately "abandoned Adarbiganon [NOTE: Azerbaijan] a lit-tle before through fear of the plague, and was off with his whole army into Assyria . . ."[48]

That Khusro was able, again and again, to probe into plague-infested territory, and then to retreat into territory that, so far, remained plague-free was partly the result of the commercial restrictions be-tween Rome and Sassanid Persia dating back to Diocletian. Confining trade with Mesopotamia to designated free-trade zones like Nisibis and Callinicum limited Persia's exposure to the pestilence, whose most important vector, the black rat, stayed close to human commercial routes.* One can freely speculate whether Khusro's relative insulation from the disease made him more aggressive in his annual invasions of Roman territory, but no one can doubt his eagerness to resume. In 544, he crossed the Euphrates once again, this time to besiege the city of Edessa.

Though Edessa is far less well remembered than other cities of the empire's eastern frontier during late antiquity, its siege offers an object lesson in the characteristic nature of these urban oases: their ability to produce commercial, spiritual, and intellectual wealth; their attraction to marauders eager to capture that wealth; and the strategies used to defend, and to assail, them. Warfare in the sixth century is less the story of decisive battles between field armies than it is of walled cities protecting their treasure.

Deciding on Edessa's greatest treasure is a matter of perspective.

---

* The dam formed by the commercial barricade was a temporary one, able to divert the river of disease only for a time. For more on the later appearances of the plague in Mesopotamia, see Chapter 13.

The Persians looked at the city, and saw gold; Justinian's subjects, on the other hand, saw the literal signature of Christ. When Procopius described Khusro's siege of Edessa as a battle "not with Justinian nor with any other man, but with the God of the Christians"[49] he was alluding to Edessa's widely known place in Christian history. In legend, at least, a prince of the city, Abgar by name, suffered from gout—so badly that he wrote to an itinerant preacher said to be performing miraculous cures somewhere in Galilee, in the province of Judea. The letter invited the preacher to leave a land where he was without respect and move to Edessa, where he would receive the respect he deserved. The preacher declined the offer, but by return letter assured the prince that gout would no longer trouble him.

Five centuries later, the letter had acquired a retrospective postscript: a promise that the city would never fall, a promise that had been inscribed on the city's gates. When Kobad attacked the city as part of his war against Anastasius in 503, the legend inspired the city's defenders sufficiently that its bishop, Areobindus, sent him a message that read "Now thou seest that the city is not thine, nor of Anastasius, but it is the city of Christ who blessed it, and it has withstood thy hosts."[50]

This city of Christ, today 'anHurfa in modern Turkey, had become, partly on the strength of its mystical connection to Jesus, home to many monasteries and the birthplace of Syriac philosophy. The most famous of churchmen to be associated with the city, Jacob Baradaeus, who gave his name to the Syrian-Jacobite church, was consecrated Bishop of Edessa sometime during the plague year of 542, and the timing was not coincidental. Jacob's sponsor, the Ghassanid ruler al-Harith ibn Jabala (Jacob's real name—both he and his king were Monophysite Christians—was Yaqub bar-Addai), presented his request to the empire's most prominent Monophysite, and, with Justinian ill with plague, its most powerful individual: Theodora. Jacob's repayment was an aggressive conversion campaign aimed at building a parallel Monophysite episcopate throughout Syria and Egypt, yet another frustration to Justinian's hopes of doctrinal reunification.[51]

Edessa was also a rich trading city, geographically well located:

Roads connected it to Armenia in the North, to Nisibis in the east, and by the original Silk Road, even to China. So when, in 544, Khusro besieged the city, he demanded one form of Edessa's wealth—fifty thousand pounds of gold, clearly a ruinous payment—while the Edessans determined that their defense would be inspired by the other form—the letter. The result was siege.

As with much else, the cinematic version of ancient siege warfare is a misleading one. Catapults hurling giant stones at walls or battalions of men throwing enormous rams at massive gates make for visually arresting images, but the reality is that, during late antiquity, the most important weapon in siegecraft was the shovel. To overcome the height disadvantage that walls present to attackers, particularly those using missile weapons who must fight gravity as well as a city's defenders, besieging armies of Justinian often put those shovels to work building ramparts to the height of the target city's walls, and even more frequently dug long tunnels underneath them, even to the point—as at the siege of the Black Sea city of Petra, in 539—of removing the walls' foundation stones. At Edessa in 544, Khusro tried to build a counter wall to the very height of the city's walls.

> On the eighth day of the siege he [Khusro] formed the design of erecting an artificial hill against the circuit wall of the city; accordingly he cut down trees in great numbers from the adjacent districts and, without removing the leaves, laid them together in a square before the wall.[52]

Realizing the danger, the city's defenders dug from the city itself underneath the mound, and "took out timbers and stones and earth and made an open space just like a chamber; then they threw in there dry trunks of trees of the kind which burn most easily, and saturated them with oil of cedar and added quantities of sulphur and bitumen."[53] When the Persians attacked, the Edessans set the enormous pile of fuel on fire.

The Persian commanders who watched their artificial mountain spewing smoke and flame out of every gap saw both the collapse of their siege engine and of the siege itself. Edessa had saved itself, and paid Khusro a token payment to speed him on his way back to Persia.

Without any perspective from his own subjects, all of whose accounts have vanished behind the seventh-century fall of Persia's empire, Khusro's historical reputation derives largely from contemporary Roman historians, and to them he presents very much a mixed bag. He is the great villain of Procopius's *Wars*; to this most conservative historian, Khusro appeared as a "man of unruly mind . . . strangely fond of innovation."[54] Agathias exhibits disdain for Khusro's supposed philosophical sophistication, essentially arguing that real sophistication was reserved to the hardy European, rather than the languid Asian. But John of Ephesus wrote that Khusro "was a prudent and wise man, and all his lifetime took pains to collect the religious books of all creeds, and read and studied them, that he might learn which were true and wise and which were foolish." He further states that Khusro esteemed the Christian Bible above other books calling it "true and wise above any other religion."[55] Other contemporaneous accounts support this in part, if only because he allowed one of his—many—wives to become a Christian.

Most significantly, Khusro represented yet another in the dizzying number of forces determining the path on which Rome's Mediterranean superstate transformed into the distinctive nations of premodern Europe. Though the cultural influence of Persia is harder to see than that of Rome—often enough, Persian assets were acquired by the empire's Arab conquerors, who preserved them—the Sassanid's political impact is clear: They are complicit, perhaps crucial, in Justinian's final failure to reunite the empire. Equally important, the extraordinarily rapid Arab conquest of Persia's empire in the seventh century is incomprehensible without understanding the brutal damage inflicted on Mesopotamia by the plague.

Khusro had been able to distance himself from the demon for four years' worth of profitable incursions into Roman territory, but the plague could only be deferred for so long. John of Ephesus and Edessa's Jacob Baradaeus record both plague and plague-caused agricultural crises in Khusro's Mesopotamian lands beginning in the year 545 and lasting for two years of "famine, plague, madness and fury."[56]

With a foothold established outside its previous eastern border, the plague spread everywhere that internal Persian trade routes carried rats: Amida (in modern Turkey, Dyarbakir) was hit by plague in 557, with a death toll of 35,000[57] and again in 558 and 559. Slightly more problematic are the likely dates of the plague's return to Antioch (which would be struck two more times before 590) and its arrival in Nisibis, but the best estimates are that the cities greeted the demon in 560. Elias of Nisibis records a plague in his city in 599, and Michael the Syrian a deadly outbreak in Palestine in 626.

By the middle of the seventh century, the Greek, Latin, and Persian accounts of Mesopotamian and Iranian plague were supplemented by far more detailed Arabic chronicles, and since Arabic actually had a very specific name for bubonic plague—*ta'un*—the record gains in precision, while losing none of the horror of previous histories. The horror survived even the end of the Sassanids, since *Y. pestis* was no respecter of different religions, and the conquest of Zoroastrianism by Islam meant little to the bacterium. Modern historians document six major recurrences from 627 onward, including the Plague of Shirawayh at Ctesiphon in 627–628; the Plague of Amwas (Emmaus, near Jerusalem), which destroyed a 25,000-man Arab army; and the Plague of Yezdigird. In 688–689, the Plague of al-Jarif—the "torrent"— attacked the city of Basra, and is said to have killed more than two hundred thousand people in three days, which, even discounting for the usual exaggeration, speaks to an enormous toll.[58]

Singularly lacking in exaggeration are the words written by Evagrius Scholasticus, who saw the plague strike Antioch four times, taking virtually his entire family with it:

> This calamity has prevailed, as I have already said, to the present time, for two and fifty years, exceeding all that have preceded it. For Philostratus [the second-century C.E. Greek sophist] expresses wonder that the pestilence which happened in his time lasted for fifteen years. The sequel is uncertain, since its course will be guided by the good pleasure of God.[59]

CHAPTER ELEVEN

"No Small Grace"

545–664

A T THE END of what had been, until then, the greatest paroxysm of violence in world history, the victorious allies of the First World War—scarcely less damaged than the Triple Alliance they had defeated—met in Versailles to determine the shape of the postwar world. In the face of a rather vague "right of self-determination" passionately advocated by U.S. President Woodrow Wilson, the premier of France, Georges Clemenceau, is reported to have asked "Must every little language have its own country?"

The Versailles conferees' combination of idealistic hopes and pragmatic *realpolitik* led, in the end, to an even more brutal convulsion, after which six nations of Europe—France, West Germany, Italy, Belgium, Luxembourg, and the Netherlands—met in 1951 to form the clumsily named European Coal and Steel Community, the precursor to what would one day be envisioned as a United States of Europe. Fifty years later, after the original community, popularly known as the Common Market, had grown into a twenty-five-member European Union, with a common currency, passport-free borders, and a rudimentary transnational legislature, a constitution was drafted by men and women who were already starting to think of themselves as citizens of such a united European state. These men and women were less quotable than Clemenceau in their attitude toward national independence, but no less dismissive. And no better at reading their historical moment; given an opportunity, in 2005, to ratify the final move toward European unity, the French and Dutch people categorically rejected it.

The history of continental Europe both before and after Justinian's last attempt to restore Roman sovereignty over it can easily be cast as a contest between these centripetal and centrifugal impulses, though calling them both "impulses" suggests a false similarity. The fundamental difference between European atomization and conglomeration is that disorganization is always easier than organization. Countering the forces of human entropy, like thermodynamic entropy, demands energy.

While many primate species form social units larger than families—primatologists call them troops—no one really knows how or when humans started down the path that led from troop to tribe to kingdom. This ignorance has failed to inhibit paleoanthropologists keen on churning out theories about it, some of them extremely persuasive. A currently popular one posits that a volcanic explosion 71,000 years ago—one far larger than the one that is thought to have changed world climate during the early 530s—led to a millennium-long ice age in which cooperation was so strongly favored that only those humans able to form larger social organizations survived.[*1]

Survived, and flourished. The best estimates suggest that human population growth and geographic range reached its nadir approximately forty thousand to seventy thousand years ago, when a global census would have numbered possibly only a few thousand individuals. *H. sapiens*'s passage through that bottleneck, whether caused by, or coincidental with, the formation of tribes, resulted in a population explosion lasting seven hundred centuries, only occasionally disrupted by ecological catastrophes such as, for example, pandemic disease. Somewhere during those centuries, tribes grew into kingdoms, even nations.

The many steps from tribe to kingdom are even more complicated than the one from troop to tribe. Human organizations at every level, from Amazonian hunter-gatherer tribes to European parliament, are

---

* The frequency with which volcanic eruptions and climate change are invoked as the motive force in human history—David Keys's *Catastrophe* is a recent example—is a subject for another book, but seems partly an artifact of the clear and lasting footprint left behind by such geological cataclysms.

assembled by a combination of self-interest and coercion, and while the former offers powerful explanations for everything from species selection to economic behavior to—perhaps—the formation of tribes, kingdoms are creations of force, frequently violent force.

If the astute application of force were the only key variable, however, the most widely spoken European language would be the Huns'. A sixth-century map of western Europe shows a bewildering assortment of protonations, each ambitious for an independent *patria*, unattainable under the hegemony of Rome.

The Goths, Alans, Lombards, Burgundians, Saxons, Germans, Thuringians, Frisians, Angles, Suevi, and Jutes (even Gibbon, who never flinched at including details, called them "tribes . . . whose obscure and uncouth names would only serve to oppress the memory and perplex the attention of the reader"[2]) were entrants in a tournament that would have few winners and dozens of losers. Retrospectively handicapping the early rounds of such a tournament, one entry seems the obvious odds-on favorite. Procopius called them "a barbarous nation, not of much consequence in the beginning."[3] As if to assert his independence from his predecessor, Agathias found them "extremely well-bred and civilized and practically the same as ourselves except for their uncouth style of dress and peculiar language."[4] Along with the rump of the Roman Empire that later generations would call Byzantine, they would dominate the Christian world for four centuries after Justinian. They were the Franks.

———

In the four centuries after Julius Caesar added the provinces of Gaul to what was not yet the rule of a Roman emperor, its residents became some of the empire's most assimilated citizens. Their attraction to Latin culture was a result of both its own merits, and because the presence of Roman legions standing watch on the Rhine against what Tacitus called "the ferocious Germans . . . who will always desire to exchange the solitude of their woods and morasses for the wealth and fertility of Gaul."[5] Gallic affinity for Rome survived the Third-Century Crisis, the civil wars of Diocletian and Constantine, and even a rather difficult fifth century, during which the western emperor Honorius exported his Visigothic problem to southern Gaul.

Though the Franks living in what are today the Low Countries were considerably less Romanized than their Gallic neighbors, they, too, occupy a place in Roman military history—several places in fact, both with and against the empire: defeated by Julian the Apostate in 359, defeating Attila at the side of Aetius in 451. Aetius's ally at the Battle of the Catalaunian Plains was the Frank's king, Merovech, who gave his name to the dynasty that would rule his people for four hundred years. But the Merovingian king who truly put the Franks on the map—more accurately, the one who truly remade the map—was Merovech's grandson Clovis.

Clovis was only fifteen years old in 481, when he succeeded to the rule of the Merovingians—some of them, anyway, since dynastic succession among the Franks divided authority among the sons of the previous ruler, and the boy's patrimony, known as the Salian kingdom, included only the territory of modern Belgium. Five years before, the last western Roman emperor had been deposed, and while the empire had exerted only nominal authority for decades, with its formal end, a number of local kinglets had emerged in the west, frequently claiming some charter from the same empire that had ruled there for five centuries. One such, Syagrius, had been ruling his share of the empire's Gallic provinces for ten years when he earned his entry in the history books as Clovis's first conquest, leaving his palace at Soissons to escape to Toulouse, where the Visigoths were still ruling the territory granted them seventy years before by Honorius.

By 496 the now thirty-year-old king had consolidated his rule over the Merovingian kingdoms, and formed marital alliances with the Burgundian king (whose niece he married himself) and the Ostrogoth ruler Theodoric the Great, to whom he married his own sister. The usual objective of such alliances is improved security, and so it was with Clovis, though not in the way anyone might have imagined: When the Frankish king faced an invasion by a German tribe—as always, the word German is not terribly illuminating; perhaps better to call them by their contemporary name, the Alemanni—who had spread from Alsace-Lorraine into the area dominated by the city of Cologne, Clovis turned to neither his Burgundian uncle nor his Ostrogoth brother-in-law, but to his wife's God.

Until then, Clovis had retained the paganism of his ancestors, but when his prayers were answered, his reaction was eerily similar to that of Constantine 184 years previously: He not only embraced Christianity himself but immediately baptized several thousand of his soldiers. The greatest of the Frankish historians, Bishop Gregory of Tours, describes Clovis's prayer

> "Jesus Christ," he said, "you who Clotild [Clovis's wife, who had somehow remained obedient to the dictates of Chalcedon while in the Arian court of her Burgundian family] maintains to be the son of the living God, you who deign to give help to those in travail and victory to those who trust in you, in faith I beg the glory of your help."[6]

and subsequent baptism thus:

> Like some new Constantine, he stepped forward to the baptismal pool, ready to wash away the scars of his old leprosy and to be cleansed in flowing water from the sordid stains which he had borne so long.[7]

Clovis was even more impatient with the theological niceties of Christology than the original Constantine; his best known theological comment is "Had I been present [at the crucifixion] at the head of my valiant Franks, I would have revenged his injuries."[8] Even so, he embraced not just Christianity, but the dictates of Chalcedon. As a result, as the fifth century turned into the sixth, and given the Monophysite sympathies of Anastasius and the Arianism of the Goths, Clovis was the only Catholic monarch from the Atlantic to the Bosporus.

Much would follow from the confessional alliance of the Frankish king and his bishops (including most of subsequent French history), but the immediate consequences were considerable enough. The two pillars of society in the territory that would one day be called France, the Frankish king and the Gallo-Roman episcopate, though warily circling around one another's prerogatives, were nonetheless allied in the goal of establishing a state with a contiguous territory, a shared religion, and a common language.

This did not imply, as it did to Justinian, a universal law. As part of Clovis's modus vivendi with his bishops, his Gallo-Roman subjects were allowed—one could apparently decide the law under which one would live by simply choosing one in the presence of a judge—to use either the Roman law, soon to be codified by Tribonian in Constantinople, or the *Lex Salica,* or Salian Code, distinguished by judicial torture, trial by combat, and *wergild.* Clovis's Frankish subjects were required to live by the Salian Code, which is frequently described as the ability to purchase a license to kill: The punishment for murder was a fine, half of which was remitted to the state. No one interested in legal perversities ever fails to notice that the Salian Code specified different fines for different victims: "A noble provincial, who was admitted to the king's table, might be legally murdered at the expense of 300 pieces . . . but the meaner Romans . . . by a trifling compensation of 100 or even fifty."[9]

While building a Frankish nation didn't mean enforcing a monolithic Frankish law, it *did* mean expulsion of anyone with other national ambitions, particularly when those ambitions were accompanied by a heretical form of Christianity. In 500, Clovis asserted his sovereignty over the Arian Burgundians, in 506 he conquered Aquitaine, and in 507, at the Battle of Poitiers, he ended the dream of a Gothic *patria* in Gaul by defeating and expelling the Visigoths of Alaric II. Clovis now ruled, from Paris, a territory that reached from the Pyrenees to the Loire, to parts as far as the Rhine. The kingdom of the Franks was well on its way to bequeathing the modern French state not only its name, but its geographic boundaries.

In 511 Clovis died, leaving only the Bretons in the north, the Burgundians in the east, and the Provençal region (blocked by Theodoric) outside the kingdom of the Franks. He was succeeded by four sons, each with a different capital, to whom he had bequeathed a powerful nation-in-formation, not least because he also left them a formal agreement with the Frankish bishops setting out a canon separating the powers between the court and the episcopate. By 532, the year of Khusro's ascension to the Persian throne, the year of the Nika Revolt in Constantinople, two of Clovis's sons, Childebert and Chlotar, had conquered and annexed Burgundy to their kingdom; by 536, seeking help against Belisarius's invading army, Vitigis, the Ostrogoth king, ceded Provence to the Franks.

The help was not quick in forthcoming. When Justinian began his invasion of the Italian peninsula, the resourceful emperor took the Franks out of the equation with a combination of diplomacy, religious solidarity—the Franks, like the Romans, were orthodox, while the Goths were Arians—and bribes.[10] Procopius quotes a possibly apocryphal letter from Justinian to the Franks in Gaul in detail: "The Goths have seized by violence Italy, *which was ours* [emphasis added] . . . and it is proper that you should join with us in waging this war, which is rendered yours as well as ours not only by the orthodox faith, which rejects the opinion of the Arians, but also because of the enmity we both feel towards the Goths."[11]

In 538, with Belisarius besieged in Rome, Clovis's grandson, Theudibert, sent the Goths some aid—though, trying to hide his support for the Arian Ostrogoths from Justinian, he decided not to use his own Frankish soldiers. Instead, he concealed his involvement by adopting the fiction that the Burgundian troops he did send were operating independently, "for it was given that the Burgundians made the journey of their own choice, and not at the order of Theudibert."[12] After the destruction of Milan, however, Theudibert led a rather large army of his own Franks into Italy, crossing the Po River at Ticino in the spring of 539, and defeating the relatively small armies of Goths and Romans (the latter under the ambitious but unlucky John the Sanguinary) in sequence. The Franks, however, were unable to avoid the dangers of a sizable army using the same water for washing and drinking, and dysentery struck them down by the thousands. Moreover, since the invasion lacked the logistical support to succeed as anything but a very large raid—the Franks had neither the capability nor the wish to engage in a war with Justinian—when Belisarius wrote Theudibert, telling him in no uncertain terms, that his choices were either to return to Gaul or to face Belisarius himself, he led his army home.

————

The first recorded appearance of the plague in Gaul occurred in Arles in 543, only a year after it decimated Constantinople. As described by Bishop Gregory, "[T]he plague raged in various parts of Gaul, causing great swellings in the groin . . ."[13] but this was merely a prelude. Thirty years later, plague broke out with a vengeance in the town of Clermont-Ferrand:

> When the plague finally began to rage, so many people were killed off throughout the whole region and the dead bodies were so numerous that it was not even possible to count them. There was such a shortage of coffins and tombstones that ten or more bodies were buried in the same grave. In Saint Peter's church alone on a single Sunday three hundred dead bodies were counted, Death came very quickly. An open sore like a snake's bite appeared in the groin or the armpit, and the man who had it soon died of its poison, breathing his last on the second or third day.[14]

After appearing in Lyons, Bourges, Chalon-sur-Saône, and Dijon, the demon then made its way to Marseilles in 588—the Frankish king, Guntram, visited a nearby village and, in Gregory's telling, anyway, prevented the disease from entering by advising the villagers to eat only barley bread, and to drink only pure water. The power of intercessionary prayer is a frequent feature of Gregory's chronicle. Two years later, the plague decimated Viviers and Avignon, "raged through other parts of Gaul, but thanks to the prayers of Saint Gall"—who led his parishioners on a Lenten march more than fifty miles long—"it claimed no victims in Clermont-Ferrand. [It was] no small grace which was able to bring it to pass that the shepherd who stayed to watch did not see his sheep devoured."[15]

The entry of the demon into the Mediterranean marked a hinge moment for both the Roman and Persian empires, though reasonable people can debate whether the pandemic caused that hinge to close, or merely eased its closing. Prior to the plague's arrival, all indications are that the population of both empires was increasing, and increasing rapidly. Evidence for such growth is found everywhere from the Balkans to Syria, with numerous and prosperous villages east of Antioch "in what is now desert."[16] The cities of Toulouse and Milan expanded their walls during the period, a sure sign of an expanding population, and—as earlier noted—the tax revenues for the empire under Anastasius grew at a remarkable rate.[17] Afterward, "it is a fact (though some historians still refuse to recognize it) that all around the Mediterranean, the cities, as they had existed in antiquity, contracted and then practically disappeared."[18] Within two years, at least 4 million of Justinian's 26 million subjects* had perished, and the imperial population continued to decline to fewer than 17 million by the end of the century.

The demon's influence on the Franks, and those portions of western Europe outside Justinian's direct authority, is equally devastating. But it is qualitatively different. The story of the Frankish kingdom's path to European dominance during the centuries after the plague is generally presented as a drama of military or doctrinal conflict, with soldiers and priests historically given the largest roles. A slight shift in perspective, however, and the leading players are no longer kings and popes, but farmers.

For reasons of both proximity and climate, the farmers of Gaul and Burgundy were peripheral to the imperial economy. The Mediterranean's two-season climate—cool and rainy from October to April, hot and dry from May to September—dictated the presence, throughout the region, of the same three core crops: wheat (and some barley), grapes, and olives. The Hellenic and Mycenaean peoples who were the first civilizations to engage in Mediterranean commerce cultivated all three, but they resided on such poor cropland that they learned, early on, how to trade two legs of the agricultural tripod, in the form of high-value-per-acre commodities like wine and olive oil, for grain, largely from Egypt.[19]

---

* The figure is from Warren Treadgold's *The History of the Byzantine State and Society* and does not include most of Italy, Africa, or Spain.

The "barbarian" territories to the north possessed a different climate and developed a far less productive agriculture. The colder, wetter soil of northern and western Europe meant wheat was much more difficult to grow there than hardier crops like oats and barley. Viniculture was restricted to the same microclimatic regions in which Riesling and Bordeaux are still produced. Olives never grew at all; northerners depended for their dietary fat upon butter and lard. The Gauls (and the Germans, the Balts, and the Saxons) accordingly produced relatively little that Rome wanted: lumber, meat, slaves, and some metals. The empire had established its Rhine and Danube boundaries for military, rather than commercial or agricultural reasons.[20] Once the expense of campaigning simultaneously against Goths in the west, Huns in the north, and Persians in the east grew, the costs of ruling those territories—primarily in Gaul and southern Germany, which could not be reinforced by sea—started to seem, in Constantinople, a poor bargain;[21] and a slow withdrawal by the empire was inevitable. During the centuries between the departure of the empire and the arrival of the plague, those lands remained the sixth-century equivalent of what would one day be called the developing world.

The plague produced a literal avalanche of demographic and population shocks, not all of them predictable. As an example, mortality among the very young, who are generally most at risk during epidemics, was actually lower than that among adults, simply because their relatively small body size offers significantly less real estate for the flea. Since only a small percentage of fleas carries the bacterium, the larger number of flea bites found on a larger body increases chances for infection.[22] For less well-understood reasons, mortality seemed to have been far higher than expected in monasteries, hundreds of which were depopulated by plague. The longest-lasting shocks, however, occurred at the agricultural base of the western European economy, and their reverberations shaped the triangular political architecture of the post-plague world. A newly Islamic Mesopotamia and North Africa—the Byzantine successors of Rome—and the Franks suffered serious declines in population, both from the direct plague-related deaths and from the declining birth rates that followed. All rebounded, with significant population increases during the ninth century and following, but only western Europe trebled its population during that period. The population explosion occurred everywhere from Britain to the Rhine to Scandinavia,[23] but the clear winners of the post-plague population race spoke neither Arabic nor German nor Greek, but Latin, thanks to one of history's least well-known agricultural revolutions.[24]

One cannot, of course, "know" this in the same way that one can know the date of the battle of Poitiers; applying economic analysis to the spotty record of commerce during late antiquity is a tricky business. However, as can be seen in a subtly reasoned 2003 paper by two development economists, Ronald Findlay of Columbia and Mats Lundahl of the University of Stockholm, it is compelling, as well, despite its reliance on a number of simplifications. Findlay and Lundahl assume that all output is agricultural, that the only inputs are labor and land, and that an equilibrium state (where mortality and fertility were in balance) existed before the advent of the plague.[25] In such a model, only farmers are technically productive; everyone else, particularly including city dwellers and the army, are not.

So, while the plague did nothing to change the inventory of farmable land, it considerably reduced the number of farmers. Too few farmers for too much land had the predictable consequence of distorting the ratio of the two "input" components of the late Roman economy. As labor got scarce relative to land, wages rose; "the rich appear to have been poorer and the poor richer."[26] Rising wages, and therefore rising prosperity, resulted in a growing population—one that could not feed itself without expanding arable acreage. As human laborers became scarcer than the land they worked, the ability of landlords—the etymology of the compound word is revealing—to keep them tied to the land was severely weakened, and the peasant, for the first time in a thousand years, was permitted to inherit, bequeath, or just depart his land.[27]

It would be centuries before the old ratios returned. By then, agrarian innovation was dead in the east, but thriving in the west and north. Agricultural yields in northern Europe had remained relatively poor until around 600. That was when the moldboard plow—the sort that turns one furrow over on another, killing weeds and permitting favored crops to prosper—made its way from China, and some enterprising farmers, driven to produce a labor-saving device in order to finesse the agrarian population decline caused by the plague,[28] made it heavy enough to cut through Gallic earth. The introduction of the plow set off a raft of other changes: Because the plow was more efficiently drawn by horses (once the horse collar was invented, anyway), farmers were obliged to plant oats as feed for them. Needing oats, they evolved a three-field crop-rotation system, replacing the "plant one and fallow the other" two-field system that had preceded it. With a three-field system, one field could be sown with wheat or rye in the fall, another with oats or barley in the spring, and the remainder left fallow, which represented a 50 percent efficiency gain. The three-field system yielded enough oats that livestock could now be raised as a money "crop," and in fact, would become one of the most important agricultural products of premodern Europe. The surplus of food and livestock attracted both wolves and wolfhounds: warriors who preyed on, and defended, the wealthy created by Europe's new, and ever more productive, farmers. More wealth paid for more formidable tools of both attack and defense, including warhorses, armor, and lances, and made the armored knight—the European version of Belisarius's heavy cavalry—the top of the military food chain.[29]

More than forty years ago, the historian Lynn White argued that the "stirrup and the plow" determined the path of modern Europe. And, to some, the effects of the plague and its sequelae don't even stop there. White also argued that the great forest-clearing project that followed the plague years—an internal Christian crusade to destroy the tree-rich sanctuaries of pagan worship[30]—opened up enough farmland to make possible the European population explosion of the following centuries.* Expanding the "frontier of cultivation" was not an

---

* William Ruddiman of the University of Virginia believes that the clear-cutting of the great European forests is the beginning of the long history of human-caused increases in atmospheric carbon dioxide, a reaction to "massive human mortality accompanying pandemics"

option available in either the built-out lands of the Mediterranean or the deserts that surrounded Mesopotamia. Only western Europe possessed the topographic advantage of a potentially arable frontier, in the form of the great European forests. Truly, the shock of the plague remade not only the political but the environmental map of Europe.[31] The great plague-caused population decline experienced in the lands north of the Alps was succeeded by a centuries-long population boom.

The inevitable result was that power flowed north from the Mediterranean, since the greatest population increase occurred among a people that paid spiritual homage to Rome, but owned fealty to a—usually—Frankish king.[32]

On the outskirts of the bishopric of Lindisfarne in Northumbria are the ruins of a small estate, known locally as Yeavering. The compound, consisting of only four buildings, was erected in the early 530s and abandoned less than fifty years later. Whether the residents of the estate died or fled is not known, but one of the buildings is a barn, complete with the characteristic bay and winnowing doors that identify it as a site for the delivery of grain.[33] With the grain came rats. And with rats, came the demon.

For six thousand years, ever since the land bridge connecting Britain with continental Europe disappeared under rising sea levels at the end of the last Ice Age, the island had been traveling a somewhat independent path. Though it experienced successive waves of migration from the continent, and traded in high-value objects like amber and precious metals, Britain was relatively detached from the Mediterranean world until Julius Caesar began its conquest in 54 B.C.E. Two centuries later, Roman domination was complete, with three legions—15,000 men—and nearly three times as many auxiliary troops enforcing Roman law from Wales to Scotland . . . or, at least, to the Tyne River, at the isthmus across which Hadrian built his wall. Those

---

("Human Disease and Global Cooling," *Scientific American,* March 2005), most especially the Plague of Justinian. An ironist can therefore take some intellectual pleasure in the notion that the origins of twenty-first-century global warming may be traced back to the dust-veil caused global cooling of the mid-sixth.

troops needed to be fed, and the grain tax imposed by the Romans in order to do so provided the impetus for a substantial grain trade that would, in the course of time, provide a dangerous lure for infectious disease.

Britain's geography did serve to insulate it from the worst excesses of the empire's Third-Century Crisis, and, as a result, it retained enough prosperity to make it a tempting target for locally grown raiders—primarily Scots. An empire already in retreat from its frontiers on the Rhine could scarcely defend even more marginal territories in Britain, so when Emperor Honorius granted British cities the unasked-for responsibility for their own defense in 410, he ended four centuries of Roman rule. A warlord named Vortigern climbed to the top of the resulting chaos, but despite his undoubted military talent, he is best remembered for the grandest strategic mistake in British history. Sometime around 430, he sent an embassy to a group of Saxons then living in what is today the Netherlands, inviting them to migrate to the west of England, there to serve as garrison troops. Like Valens inviting the Goths to cross the Danube, he lived to regret his mistake, though not to correct it. The subsequent civil war was ended at the Battle of Badon Hill (*Mons Badonicus*),* but by then Britain was firmly divided along an east-west axis: the Romano-Celtic west, with a long history of seaborne trade (the wealth of the west dates back to the same kings of Wessex who built Stonehenge) and the Saxon east. The combination of a widespread grain-producing infrastructure and a maritime-trading culture was to prove unhealthy for Britain. As, indeed, it did for Ireland.

By the sixth century, despite even greater geographical isolation than Britain, Ireland had been an active commercial partner with Mediterranean shippers for centuries, trading pottery, cloth, and—of course—grain.[34] Its position on a popular commercial sea lane placed it in the path of the demon, and it is, therefore, unsurprising that Ireland started to experience bouts of epidemic disease as early as the mid-

---

* By a Roman-British officer named Ambrosius Aurelianus, a popular choice for the source of the Arthurian story cycle.

540s. Even so, it cannot be determined whether the disease was the bubonic plague—descriptions of "the great mortality called 'blefed' according to the Annals of Tigernach"[35] do not mention buboes.

The farther the demon ranged from the eastern Mediterranean, the less clear its outlines in the historical record. The appearance of the plague in Ireland and Britain in the 540s has been obscured by its recurrence 120 years later; most of what is known of the travels of *Y. pestis* across the English Channel and the Irish Sea concerns the plague of 664. In some Irish chronicles, the Irish plague appears as the *Buidhe Conaill*, best translated as "the yellow end," though the term is more confusing than enlightening, since jaundice is more likely to be a symptom of malaria than bubonic plague. The records of the demon's passage to the British Isles are notoriously obscure; while some estimates are that a third of the Irish population was infected, most of the history is derived from anecdote. Identified victims include St. Aileran of the Wisdom, and St. Ultan, a bishop who survived the plague and founded an orphanage for the children of those who didn't.[36]

The story of England and the seventh-century plague is known to us largely through the work of a single man. He was born and raised in the enormously wealthy monastery of Wearmouth and Jarrow in Northumbria, and, due to both his native aptitude and his access to the splendid library that the monastery's founder had accumulated, was very likely the most educated man of his world, the author of dozens of writings on everything from geography to musicology. But the best-remembered work of the monk called Bede—he would not be called "Venerable" during his lifetime—is his *Ecclesiastical History of the English People*, which he wrote in 731, only a few years before his own death.

About the pestilence, Bede wrote, "[It] raged far and wide with fierce destruction . . . [and] ravaged Britain and Ireland with cruel devastation."[37] By then, the demon was appearing only sporadically in the Mediterranean, and was extinct in Ireland; but for two years, until 666, plague ravaged England. Of Britain's eight bishops in 664, four died of the plague, as did the kings of Kent and Northumbria. It returned (as it did so often in the Mediterranean and Mesopotamia) twenty years later, this time in 684, where it devastated Northumbria again, destroying virtually the entire Lindisfarne congregation. Adomnan, the Abbott of Iona, wrote in 697 of "the great mortality which twice in our time has ravaged a large part of the world."[38]

The two churchmen, monk and abbott, are reliable reporters. No one who reads Bede fails to notice the measured tone of his *History*, for while his perspective is religious, it is not apocalyptic. His descriptions of the disease are not quite up to the level of Procopius (or even Gregory), but he does notice that Saint Cuthbert showed a "swelling on his thigh."[39] Even so, the lack of corroboration from either gravesites or funerary documents has demanded a good deal of inferential research by modern historians seeking to establish the range and virulence of the sixth- and seventh-century plagues. The fact that monasteries— including Whitby and Lindisfarne—were largely located on coasts or estuaries, and probably offered multiple entry points for the shipborne rats, may have reflected their "preeminent vulnerability to infection"[40] that, in the words of Procopius, "always took its start from the coast." And while the ubiquity of the grain trade—for generations, kings and lords took their rents in the form of grain—cannot prove the existence of the plague, it offers an intriguing clue nonetheless.

The wine and oil merchants who inaugurated the Mediterranean grain trade established a wide-open pattern of seaborne commerce, even after the fourth-century empire chartered a hereditary shipping guild—in Latin, *corpora naviculariorum*—to supplement the thousand galleys owned by the state exclusively for the transport of African grain to Rome.[41] The result was a Mediterranean trading system expansive enough to embrace Britain and Ireland, and to provide a network for transmission of disease.[42] In the end, the magnitude of the *mortalitas magna* in Ireland and Britain cannot be known, but its presence is certain. The cemetery at Camerton in Somersetshire is home to 115 skeletons, including 40 children, buried hurriedly in shallow graves:[43] victims—like their brethren in Gaul, Africa, Mesopotamia, and Italy— of the world's first, and by some measures deadliest, pandemic.

CHAPTER TWELVE

"A Thread You Cannot Unravel"

548–558

THE SUBTERFUGE BY which Belisarius, in 540, secured the surrender of the Ostrogoths in Ravenna ended a battle, but not the war. In part because of the nature of his victory—via the pretense of accepting a crown of his own—Justinian's pathological suspicions were aroused to the point of recalling his most successful (and therefore most threatening) general to Constantinople.

The emperor's decision not only removed his strongest military asset from the field but also left that field without *any* supreme commander. The consequences were predictable; while Belisarius was able to deliver victories with even the small army provided by his parsimonious sovereign, his successors were not, and the mopping-up operations needed to secure the imperial victory quickly collapsed. In their place, the occupying army established garrisons in Ravenna, Florence, and Rome, and around the Venetian lagoon—the city proper would not be founded until 558—largely devoted to supporting imperial tax collectors sent to make the peninsula pay for its reconquest, while skimming a fair share of plunder themselves. Meanwhile, the Gothic armies that had returned home after the 540 surrender started to reconstitute themselves north of the Po. Their leader, Vitigis's successor Hildebad, enjoyed some victories in an insurgent campaign against Justinian's occupying troops but was assassinated by one of his own bodyguards only months after taking power. His replacement, Eraric, tried to sell out to Justinian on the sly, but when his plans came to light, in the fall of 541, he in turn was killed and replaced by Hildebad's nephew, a young and charismatic soldier named Totila.

Totila wasted little time in making the imperial troops bleed. In the spring of 542, Totila, having defeated a Roman punitive expedition at Verona, crossed the Po and advanced into Tuscany; the Roman garrisons that might have confronted him remained behind the walls of their fortified cities, which gave the new Ostrogoth leader both mobility and initiative. Taking advantage of both, Totila bypassed both central Italy and Rome itself, marching south to occupy Apulia—Italy's boot—and besiege Naples, where, after several inept and unsuccessful attempts to lift the siege, the city surrendered in the spring of 543. Totila spent the next year consolidating his gains while imperial troops, far from stopping him, stepped up their rate of plunder. With the Ostrogoths effectively in control of the northern and southern thirds of the peninsula, Totila's insurgency had become a full-fledged rebellion, whose next target was the city of Rome. His chosen weapon, however, was not a battering ram or a siege tower, but a letter to the city's noble families. The letter, which reminded the Romans of the prosperity they had enjoyed under the rule of Theodoric and compared it to their relative immiseration ever since sovereignty had been returned to Justinian, was received so well by the Romans that, in the spring of 544, Totila marched north at the head of a token army, expecting to be greeted in the Eternal City as a liberator rather than a conqueror.

The attempt to conquer Rome bloodlessly added luster to Totila's reputation, though not in a completely unexpected manner. During his successful siege of Naples, the Goth treated the population with notable courtesy, allowing the Neapolitans to keep their treasure, even letting members of senatorial families depart freely, a conscious imitation of the technique Belisarius used to win the sympathy of locals in campaigns ranging from North Africa to the Italian Alps. Totila's unlikely blending of the skills of a clever diplomat and a barbarian warrior likely explains the high regard in which he was held by medieval Germans, who adopted him as one of their great Romantic heroes (frequently under the name of Baudila). The problematic truth of this image—Totila's leniency toward the Neapolitans is better remembered than his cruelty toward, for example, non-Arian priests[1]—is less important than its resonance. To Germans who celebrated the Ostrogoths as their Golden Age forebears, Totila/Baudila was another Theodoric, and the latter was such an appealing ancestor that he appears—under the name Dietrich von Bern—as the hero of the anonymous thirteenth-century epic, and source of Wagner's great opera, the *Niebelungenlied.*

————

In Constantinople, the march of Totila toward Rome did not pass unremarked. Later the same year, faced with the difficulty of consolidating what had seemed, only a few years previously, a conquered Italy, Justinian responded to the latest Gothic threat with what had become practically a reflex: He sent Belisarius. And again, he sent little else; the entire expeditionary force numbered only four thousand, and even they were poorly armed and untrained recruits. Arriving in Italy at the head of this unimpressive force, only to discover that most of the remaining imperial garrison troops were either in hiding or had deserted to the enemy, the general drafted a letter to the emperor, demanding help: "If . . . it was sufficient to send Belisarius to Italy, then our job is done. If, however, it is thy will to overcome thy foes, it is needful that my spearman and guards"—the general's private retainers, his *comitatus*—"and a large force of Huns be sent immediately."*[2]

Belisarius's recruiting shortfalls were not entirely due to desertions. Justinian's economy had already been under some severe strains before the arrival of the plague. Though the raison d'etre for the emperor's western conquests had never been economic, the Persian wars and the North African and Italian expeditions had eaten through most of the 320,000 pounds of gold inherited from Anastasius, and—in a reminder that conquests rarely pay for themselves—with the exception of some of the North African Vandal territories, the administration of the newly conquered territories was far more costly than the tax revenues that they generated.[3]

---

* Belisarius made an unfortunate choice of messengers to carry his plea: John the Sanguinary. Despite promising a rapid return with the needed troops, the matrimonially minded soldier—recall his distracting courtship of Matasuntha while armies formed outside of Ravenna—decided to spend a month wooing and finally marrying another royal, the daughter of Justinian's cousin Germanus.

War is hard on an economy; pandemic disease is even worse. The best evidence for the demon's economic damage is Justinian's response: In March 544, as Khusro was besieging Edessa and Totila was marching north to Rome, Justinian issued an edict "on the regulation of skilled labor" in which he froze prices and wages at their preplague level:

> [P]ursuant to the chastening that we have received in the benevolence of our Lord God [i.e., the plague] persons . . . have abandoned themselves to avarice and demand double and triple prices and wages that are contrary to the custom prevalent from antiquity, although these persons should rather have been improved by this calamity. It is therefore our decision to forbid such covetous greed on the part of everyone. . . . In the future no businessman, workman, or artisan in any occupation, trade, or agricultural pursuit shall dare to charge a higher price or wage than that of the custom prevalent from antiquity.[4]

As with all such wage and price freezes, Justinian's edict served mostly to exaggerate the problems they were intended to solve. Shortages of fuel, food, and—most relevantly for an empire now fighting a two-front war—the necessaries for battle, including weapons and men, grew even more acute.

Far fewer such shortages affected the Ostrogoths, which was fortunate for Totila since, despite all his attempts to suborn the garrison of Rome, in late 546, he was still compelled to besiege the city. Meanwhile Belisarius—as always on campaign, accompanied by Antonina—attempted to raise the siege from his base at Portus, on the Tiber downstream from Rome. The weak point to any siege of a riverine city is, of course, its river, and Belisarius immediately set about building fireships to destroy the booms that Totila's engineers had used to block the Tiber's access to the city. The fireships worked brilliantly, and, but for a misadventure that forced Belisarius to abandon the attack in order to rescue Antonina, might have broken the siege. In the event, he failed, and on December 17, 546, Rome surrendered to Totila . . . a negligible event, since by the time the garrison of three thousand men escaped, only five hundred civilians remained in the entire Eternal City, now truly a ghost town.[5]

By spring, Totila had tired of his depopulated conquest (upon entering the city, he had started to destroy many of the city's monuments, but had been dissuaded by a letter from Belisarius, still encamped in Portus, reminding his adversary that "an insult to the monuments would be considered a great crime against all men of all time"[6]) and departed for the north. Now thrice deserted—once by its garrison, once by its civilian population, and now by its Gothic occupiers, Rome was reentered by Belisarius and his troops.

The lesson of the siege was acutely obvious: While he might be able to harass Totila, Belisarius could not defeat him with the limited resources that remained after ten years of war and—more relevantly—four years of plague. In a last-ditch attempt to persuade Constantinople of his need, Belisarius decided to send Antonina east, to plead not with Justinian but with Theodora.

———

More than 150 years ago, the English historian C. E. Mallett could write of Theodora, "There are few stranger episodes in literary history than the fate of this celebrated empress,"[7] but the empress's remarkable life has always threatened to overshadow her historical importance. She was venerated for a thousand years, and only with the discovery of the *Anekdota* by a Vatican priest in 1683 was that image smudged by scandal. Procopius's salacious image of Theodora as a sybaritic gourmet with a voracious sexual appetite, spending hours in her bath, napping every afternoon, traveling to her Black Sea vacation palace on barges containing thousands of retainers, complicated the fascination she exerted thereafter (and still does today) but did nothing to diminish it. Women of profound and sincere faith who are also proto-feminist icons have never been common.

But even were Theodora's life less irresistible as a saint-and-sinner drama—literally, since, like her husband, Theodora is a saint of the Eastern Orthodox Church; the façade of the Church of Sts. Sergius and Bacchus described her as "God-crowned Theodora, whose mind is adorned with piety and whose constant toil lies in unsparing efforts to nourish the destitute"[8]—she would still demand attention because of her complicity in the divorce between eastern and western Christianity.

From the time she married Justinian, Theodora made her cause the survival of the Monophysitism to which she had pledged loyalty during her time in Alexandria. Justinian may have sought a doctrinal rapprochement between the Monophysites and Chalcedonian/Catholic orthodoxy, but Theodora's goals were less lofty: She wanted Monophysite hands operating the levers of the Church's authority. Her attempt to install Monophysite bishops in important eastern sees, up to and including Constantinople, failed—Pope Agapetus persuaded Justinian to replace them with more orthodox choices—but so did Justinian's hopes of an east-west reconciliation. To be sure, the divorce between them might have proceeded in any case. An emperor and a pope could not share the ecclesiastical leadership—this is a theme that would recur a dozen different times in European history, from Frederick II to Henry VIII. So once the eastern churches, with Theodora's support, asserted doctrinal independence and the western Chalcedonian/Catholics gained political independence (a direct consequence of Justinian's inability to consolidate his victories in Italy and Spain), the future direction of the Church—Roman Catholicism in the west, various Eastern Orthodox rites in the east—was impossible to change.

Given her inclinations and enormous influence on matters religious, one can scarcely be surprised that the most reproduced image of Theodora appears in a church. The only surprise, in fact, is that the church in question is not in Constantinople.

The Church of San Vitale is one of Ravenna's best-known architectural landmarks; it is surely the most visited. Like Theodoric's tomb and the two baptisteries, it is a relatively small octagonal building built on a typical central plan, but the church's artistic significance lies not in its structure, but in its decoration. The great mosaics of Justinian and Theodora face each other across the apse of the church, and are telling evidence not merely of the empress's great beauty—huge eyes, wavy hair, an expressive mouth—but of the imperial philosophy sometimes called caesaropapism: the ascendance of emperor over pope, since the former is Christ's regent.[9] The mosaic Justinian is accompanied by twelve men, and even sports a halo surrounding his crown. Theodora's dress features an image of the Three Magi. Both carry offerings. As a historian put it sixty years ago, "[N]o other work of art . . . conveys the spirit of Byzantium with so much eloquence as do these two mosaics."[10] Begun in 526, a year before Justinian took the throne, the church was dedicated the year that Antonina set sail for Constantinople on her mission. She arrived just in time to bid her oldest friend farewell.

On June 28, 548, Theodora died, most probably of cancer. For days, the bearkeeper's daughter lay in state in the imperial palace, dressed in purple on a gold catafalque. The people filing past in review included the patriarch, senators, priests—one of them, Anthimus, was the long-ago deposed Bishop of Constantinople, whom Theodora had been hiding in the Hormisdas palace ever since[11]—and the leading members of every aristocratic family in Constantinople. Side by side were dozens of former prostitutes and actresses who had been housed and clothed by the empress who had never forgotten them in life. The last to approach was Justinian himself, who broke down after placing an enormous necklace around his wife's neck.

Twenty-one years earlier, the newly crowned emperor and empress left the old Theodosian church that had then occupied the site of the Hagia Sophia and led a jubilant crowd to the Hippodrome. The last procession Theodora would lead as empress walked slowly down the Mese, past the Forum of Constantine, to the Church of the Holy Apostles—Constantine's burial site, and the model for Venice's San Marco. As the body entered the nave of the church, the mandator cried: "Enter into thy rest, O Empress. The King of Kings and Lord of Lords calleth thee." Following the funeral mass, the empress was interred in an enormous sarcophagus of green porphyry,[12] and the entire city entered into a state of mourning, with no one grieving more than the emperor. Flavius Corippus, a North African scholar, wrote "The old man no longer cared for anything; his spirit was in heaven."[13]

While those who loved Theodora grieved at her death, others cheered. One was her old enemy John the Cappadocian, who was finally allowed to end his exile and return to Constantinople.

Justinian's gratitude and affection had saved John from execution or even imprisonment after the Cappadocian's entrapment into treason by Theodora and Antonina. The emperor did, however, banish his prime minister to Cyzicus, and strip him of his titles and most of his wealth, though John retained enough of the latter to live most comfortably in exile. This had been adequate punishment for Justinian, but not for Theodora, whose ruthless hand can be seen in the next chapters of John's descent. With his never-failing ability to create enemies, John nurtured a dispute with Cyzicus's local bishop, a man named Eusebius. When the bishop was mysteriously killed, John was immediately suspected, and while the senators sent to Cyzicus to investigate found his guilt unproven, the one time Praetorian Prefect was nonetheless stripped, whipped down the main street like a common thief, and sent by ship to Egypt where, legend has it, he was forced to beg in the seaports in which the ship stopped en route. Once in Egypt, he was imprisoned.

Even this was not enough for Theodora, who tried to extract a confession from two men accused of Eusebius's murder—a confession that would name John. In this, she failed, despite bribes, coercion, and torture. She was, however, able to keep the onetime prefect in a purgatory that would end only in death. John, unable to outmaneuver the empress, outlived her instead; and he returned to Constantinople a free man, though a poor one, and a priest at that.

Theodora's vendetta had claimed victims other than John, including, perhaps, Justinian himself, who was deprived of the services of one of his most formidable lieutenants. "Few of his successors won any sort of reputation, whether for good or ill, and none had John's opportunity to apply consistent, forceful direction to the adminstrative processes that supported the military effort."[14] The most notable achievement by any of the subsequent Praetorian Prefects is probably the inflation of the imperial currency by the otherwise forgotten Peter Barsymes.*

The actions of another enemy who cheered Theodora's death had a rather longer tail, one that led, circuitously, to the final conquest of the Ostrogoths. The antagonist in question was Artabanes, a general from Persian-controlled Armenia who had been responsible for the death, in 538, of Justinian's general—and Theodora's brother-in-law—Sittas.

In the course of events, Artabanes transferred his allegiance to Rome, and was sent to Libya in 543, a territory suffering both from a widespread insurgency and the arrival of the plague: "The whole earth was ravaged by an epidemic and a [swelling] of the groins, so that the better part of the people perished."[15] One victim of the insurgency was a senator named Areobindus, whose widow's charms proved irresistible to Artabanes. His courtship included avenging her husband's murder, but, in order to marry, the Armenian needed more than the widow's gratitude. Since Praejecta was not merely wealthy and connected to the aristocracy but also the niece of the emperor himself, Artabanes required imperial approval to marry. In 546, however, when the Armenian pled his case before Justinian and Theodora, the empress not only turned down her brother-in-law's killer, but insisted that he return to his wife.

After the death of Theodora, Artabanes left his wife; but by that time, Praejecta had married someone else, which turned his resentment into treason. The Armenian approached Justin, the oldest son of Justinian's general Germanus—trying to foment a rebellion.

In 549, in an eerie replay of the fall of John the Cappadocian, Justin went to his father, who in turn reported the plot to the same captain of the guard, Marcellus, who had been recruited by Theodora to

---

* The "lightweight" gold coins of the Barsymes prefecture were not without importance of their own. Kenneth Harl of Tulane University has calculated that minting coins with a purity of 20 carat instead of 24 could have produced as much as 168,000 *solidi* for the imperial treasury: more than 2,000 pounds of gold. This was nearly the annual payment owed by the peace treaty with Khusro.

witness John's confession of treachery eight years before. As reported
to Justinian, however, the plan was not merely to kill Justinian, but to
place Belisarius on the throne. Though the plotters had apparently involved the general without his knowledge, Justinian calculated that so
long as Belisarius was known as the most successful military leader in
the empire, he would prove a temptation to rebellion. As a result, and
in a perverse conclusion to Antonina's mission, Belisarius was recalled
from Italy, and—in legend, at least—reduced to penury, and effectively placed under house arrest.

The removal of Belisarius may have quieted Justinian's fears of
usurpation, but did nothing to further the defeat of Totila. For that,
another general was needed, and Justinian tried quite a few, including,
in order, a former functionary of Theodoric named Petrus Marcellinus
Liberius; the traitor Artabanes who, having impeached Belisarius, was
now apparently returned to imperial favor; and the emperor's cousin
Germanus.

In 550, Justinian selected Germanus, a choice that was only partly
due to his undoubted skills as a soldier. That same year, Germanus, by
now a widower, had married Vitigis's widow—and, more relevantly,
Theodoric's granddaughter—Matasuntha.* The newly appointed
general clearly shared his emperor's belief that the Goths would think
twice before fighting an army that included Theodoric's granddaughter, since he planned to bring her to Italy with him. However, after assembling his army at Serdica (modern-day Sofia), he fell ill and died.

And there the army remained, delayed both by the need for a new
commander and by yet another people-in-arms who would play an
enormous role in the birth of Europe: the Slavs.

Little is known of the Slavs before Justinian's expeditionary force
encountered them in 550. The name is inconsistently applied throughout the sixth century; they were sometimes called Sclavenes, sometimes Bulgars or Bulgarians. But by whatever name, they followed in
the slipstream of the great Gothic migrations by moving northward

---

* John the Sanguinary had by this time married Germanus's daughter Justina, thus making his
onetime fiancée his mother-in-law.

toward the Dnieper, west to the Elbe, and south into the empire. John of Ephesus, in 581, wrote of the first appearance of what he termed "an accursed people, called Slavonians, [who] overran the whole of Greece, and the country of the Thessalonians, and all Thrace, and captured the cities, and took numerous forts, and devastated and burnt, and reduced the people to slavery, and made themselves masters of the whole country, and settled in it by main force, and dwelt in it as though it had been their own. . . ."[16] In the slightly more even-handed eighth-century *Strategikon* of the emperor Maurice they are described as "independent . . . populous and hardy . . . kind and hospitable to travelers" and yet "completely faithless . . . always at odds with each other [so that] their common hostility will not make them united or bring them together under one ruler."[17]

Though their impact would be as significant, and even longer lasting, than that of the Goths, the immediate consequence of the Slavic crossing of the Danube in 550 was to delay Justinian's latest attempt at resolving his Italian reconquest. Having lost his greatest general to his own paranoia, and Germanus—very likely his second greatest—to death, Justinian needed his unerring eye for talent as never before. It did not fail him. The man the emperor named to supreme command in April of 551 was then probably already in his mid-seventies, and had to be recalled from the retirement home he had built in Cappadocia. According to Procopius, the reason for the choice was a prophecy that the ruler of Rome would be defeated by a eunuch, but more probably Justinian, as he had so many times before, found the perfect combination of talent and loyalty in Narses.

Though he had acquitted himself well in his earlier tour of duty in Italy, Narses had never seemed to have Belisarius's flair, and he was by now ten years older. Though speculation about the source of the military aptitude Narses would henceforth exhibit is common—the great military historian Sir Basil Liddell Hart suggests that he was book taught[18]—neither his library nor his earlier generalship would have taught him much about grand strategy, which in this case meant war at sea.

From the time of the Punic Wars, Mediterranean conflict had

always favored the better navy, since the same technology that privileged sea trade over land—and transported plague-ridden rats—offered a huge advantage to the side that could dominate the sea; one naval historian even goes so far as to compliment Justinian's choice of beginning his program of reconquest by attacking the Vandals—the only opponent with real naval capacity.[19]

Totila understood this better than any of Justinian's adversaries. Not only did the Ostrogoth invest in dozens of warships but he did so at a moment when the empire was experiencing tremendous losses in merchant shipping because of the plague. Before Narses could bring the Goths to conclusion, he needed to be able to transport an army to the Italian peninsula, and to supply it once there. That meant control of the Adriatic, which could now only be secured by decisive battle.

The immediate objective of the naval battle of Sena Gallica was the relief of the Roman garrison in Ancona, which was simultaneously besieged by Totila's army, and blockaded by his navy. Ancona was strategically important as the city Narses had identified as the staging area for his expeditionary force—in the words of John, still awaiting Narses's arrival, "War depends for its decision in large measure upon the commissary . . . for valor cannot dwell together with hunger."[20] To break the blockade by Totila's fleet, Narses's galleys rowed their way into a line in order to face, bow to bow, sixty-one of Totila's, firing arrows and other missiles. But while the Goths had been on blockade duty, Narses's sailors had spent months drilling in the techniques of close combat, and, in consequence, his galleys were able to stay close enough to support one another but not so close as to hinder coordinated movement. One Goth ship after another was isolated by imperial archers, and rammed by Roman galleys. Nearly fifty Gothic ships were sunk or taken; the surviving eleven were beached by their commanders, and set on fire to keep them out of Narses's hands. When their crews marched to join Ancona's besieging army, their presence was more damaging than helpful to the Goth cause: The news of the defeat convinced the entire army to retreat.[21]

The defeat of Totila's galleys had removed the possibility of harassment by sea, but Narses remained unable to find sufficient sailors to carry his army to Italy. The plague always attacked some professions more than others, usually for predictable reasons: large numbers of people living in close confinement; proximity to rats. Like monks, who suffered a higher-than-average exposure to the demon because of the close quarters of monastic life, sailors were likewise decimated by the disease, and their plague-caused scarcity obliged Narses's army to travel overland to Padua before turning south into the heart of the peninsula.

The size of the army that Narses brought into Italy reflected Justinian's determination to send, for the first time, a force large enough to achieve its objectives; its composition reflected its commanding general's experience as an emissary of Justinian to a variety of barbarian peoples: three thousand Herulian cavalry, two thousand Huns, four thousand Persian deserters, between ten and fifteen thousand imperial regulars largely from Thrace and Illyria, and nearly six thousand Lombards.

The Lombards, one of the tribes that formed the barbarian confederation known as the Suevi (whose most notable contribution to the history of late antiquity was the Roman generalissimo Ricimer) were, in some readings, named for their predilection for wearing their beards long: *Langobardi*. By the time of Justinian's reconquista of Italy, this relatively peaceful tribe had migrated south to occupy most of present-day Austria, with the emperor's explicit permission. By 546, their peacefulness was long gone, a result of the militaristic proto-feudal culture established by their ambitious king, Audoin, with dukes, counts, and the like commanding and arming bands of retainers. Armies of conquest, like any good tool, need to be used regularly, and Audoin spent twenty years honing his forces in battles against the Gepids. By the time Narses hired Lombard mercenaries for his decisive attack on Totila, they had developed a well-deserved reputation for ferocity.

The route of Narses's coalition was the only one permitted by geography—other routes were easily flooded by damming the Po and the Adige, both of which Totila did—and what were just beginning to become the borders of future European states: The right flank of Narses's march was held by the Franks, who were disinclined to permit an army containing six thousand of their Lombard enemies to pass. But with Ancona in hand, and his army making its way through the narrow door to Italy, Narses could proceed to force a decisive battle on land against his opponent, and with nothing less in mind, he led nearly 30,000 soldiers south to Rimini, and from there to Taginae, where Totila, at the head of 15,000 troops, awaited him.

Given the significance of the battle of Taginae in subsequent history, remarkably little is certain about it. Debate continues regarding the route taken by Narses (all agree that he was obliged to leave the coast road and to avoid an easily defensible tunnel, itself a marvel of first-century Roman engineering) the actual site of the battle, or even its name, which was given by Procopius (and dozens of subsequent historians) as Busta Gallorum, commemorating a third-century B.C.E. battle by Roman troops against the Gauls.

The battle, by whatever name, began with the two generals trying to outfox each other. Narses offered Totila the choice of either surrendering or picking a date for combat (the Battle of Soissons, in 486, featured a similar offer from Clovis to Syagrius[22]). Totila responded by naming a date eight days hence. Neither won any points for either honesty or credulity; and the following day both armies were drawn up for battle, with many soldiers on both sides surprised to see their opposite numbers across the fields of Taginae.

The least surprised man was probably Narses himself, whose plan all along had been to draw his much younger opponent into the attack, where he could be surrounded and counterpunched to death. In the words of Liddell Hart, the Armenian's success in outflanking his enemy by marching into the heart of Italy and choosing the site of battle showed a gift for the strategic offensive; his actions in the decisive battle of Taginae exhibited an equally well-honed talent for a tactical defense.[23] He dismounted a large number of his allied cavalry, armed them with lances, and placed them in the center of his line. On either side of his temporary phalanx, he placed foot archers, and on either side of them, the remaining cavalry. The left wing of the cavalry was led by Narses himself.

Facing them was Totila's army, also disposed in line, but with ten thousand fewer soldiers, he was overlapped on each flank by Narses's forces. Totila now understood the importance of stalling while a much-needed two-thousand-strong Goth cavalry raced to arrive at the battlefield. His next maneuvers, therefore, were almost certainly nothing more than a calculated delaying tactic, but they are generally seen to carry a sturdier metaphorical weight, symbolizing, perhaps, the eternal conflict between barbarism and civilization.

For, as the morning wore on—Narses didn't want to give up his tactical defense, and the Goths weren't yet ready to attack—Totila himself rode into the no-man's land separating the two armies; and, in all fairness, it was perhaps not merely a desire to delay the inevitable that persuaded the Goth in his display of trick riding, throwing his spear up in the air, spinning it like a baton, and catching it. One suspects that he intended it as both a morale-builder for his own outnumbered troops as well as a pointed reminder that he was considerably younger and more energetic than his septuagenarian opponent. To Procopius and uncounted numbers of other chroniclers, however, the scene also represents the classic image of the barbarian warrior: strong, skilled, and noble; facing and, ultimately, losing to an army of professional soldiers.

Totila certainly offered considerable evidence in favor of such a metaphorical reading of his behavior. Given the strong defensive position established by Narses and his obvious preference for defense, one can only wonder why Totila played into Narses's hands so cooperatively, unless he was overconscious of the need for upholding a warrior's honor. First, he ordered a winner-take-all cavalry charge right into the teeth of the Roman preparations, then (for reasons that remain unexplained) commanded them to use only their lances. No swords, and—more tellingly—no bows, the weapon of cowards.

It took medieval Europe eight centuries to figure out what Narses seems to have intuited on a single day: Dismounted pikemen supported by missile artillery are more than a match for heavy cavalry.* As with the French knights at Agincourt and Crecy, like Lord Raglan's cavalry at Balaclava, the carnage was horrible. In the initial charge alone, more than six thousand Goths became casualties.[24]

The battle ended with that charge. So did Totila, though the details are blurred by there being several versions of his death. In one, the Goth leader died in the first flight of Roman arrows, perhaps explaining the questionable tactics that his troops exhibited the rest of that long day. In another, he was killed by mistake by one of his own lieutenants, and dragged to a village where he was buried in a secret grave. A Goth woman, however, observed the burial, and revealed it to Narses's troops, who exhumed the body and sent Totila's clothes and crown to Justinian.

Narses now felt confident enough to dismiss his Lombard allies,† whose ferocity no longer made up for, in the words of J. B. Bury, their taste for arson and rape.[25] Even without them, he had little difficulty in taking Rome, defended now by such a small garrison that the only Gothic resistance came from the same building—Hadrian's tomb, now the Castel Sant'Angelo—where Belisarius's soldiers had turned back a Gothic attack by using the destroyed statuary as missiles. It was the fifth and last time that the city would change hands during the war, and, perhaps, the most significant. "The Rome that Belisarius delivered was still the Rome of the Caesars; the Rome that Narses entered sixteen years later was already the Rome of the Popes."[26]

---

* Likewise, the medieval knight was prefigured by Persia's heavy cavalry, as translated by Belisarius and other Roman generals. The most vivid image of medieval jousting—knights charging at one another with couched lances—appears in Sassanian art from as early as the third-century; the rock carving of Naqsh-I-Rustem shows Persian and Roman horsemen charging at one another in precisely this manner. (Edith Porada, *The Art of Sassanians: Iranian Visual Arts*, Iran Chamber Society online article, 2001–2005.)

† He did not, however, forget them. Years later, the empress Sophia, wife to Justinian's successor Justin, insulted the now retired general by telling him—in legend, at least—that as a eunuch, he was only fit to spin among maids. He responded with the rather oblique threat that he would "spin a thread you cannot unravel" and did so, inviting the Lombard king Alboin and his onetime Lombard allies to return to Italy and take what they liked. True or not, the Lombard invasion was the most destructive in the long history of the Italian peninsula. (Fauber, *Narses: Hammer of the Goths*.)

One final battle remained. Totila's successor, Teias, had escaped Taginae and was able to assemble seven thousand Goth troops for a last stand against the empire at Mons Lactarius, just north of the Amalfi peninsula, on the plain at the base of Mount Vesuvius, where he confronted fifteen thousand of Narses's soldiers in March 553. Though a description of the battle that features Teias catching a dozen javelins on three separate shields, and finally being killed as he was reaching for a fourth shield, is almost certainly fanciful, after two days of vicious hand-to-hand combat, Teias indeed perished. The Ostrogoth kingdom that had ruled Italy for a century had vanished forever.

Their Gothic cousins further west, in Spain, fared a bit better. After the death of Theodoric in 526, the Visigothic kingdom was once again ruled by a member of the Balthi dynasty, though after the arrival of the plague in 543[27] its independence was shaky. Not only were three successive Visigothic kings assassinated—Amalaric in 533, Theudis in 548, and Theodegisel in 550—but revolts became endemic. In 551, a Visigothic noble named Athanagild rebelled against King Avila and called on Justinian to back his play. In June of 552, a Roman army landed near Malaga, and within weeks defeated Avila at the Battle of Seville. At the same time, the rival kingdom of the Suevi, ruling what is now Galicia, converted from the Arianism it had adopted in the early fifth century to Catholicism, thus surrounding the Arian Visigoths. Another of Justinian's armies seized Cartagena in the spring of 555, and there they stayed, dominating Spain's Mediterranean coast for the next seventy years.

As with so many of Justinian's undertakings, the Spanish conquest echoed for centuries. For while the Visigoths remained in control of the remainder of Spain, their uncomfortable position as an Arian pocket whose only access to the Mediterranean was blocked by a determinedly Catholic province led to their mass conversion in 590. And the echoes didn't stop even then. After the Arab conquest in 711, the successors of Alaric's Arian Visigoths would form the seed of the Catholic, Spanish-speaking Aragon and Castile that Ferdinand and Isabella would lead on their own *reconquista* and dominance of European politics for centuries.

Narses's victory over the Goths was unequivocal, but its consequences were not. Those who regarded the Goths as Arian heretics welcomed their downfall, but later historians speculated whether Justinian's strategy, however well implemented by Narses, succeeded only in beggaring the empire. One of the latter, H. A. L. Fisher, argued that Narses's defeat of Totila determined European history for a millennium. As the twentieth-century philosopher Sidney Hook wrote, "error or blessing, the act was fateful for the history of Europe."[28]

Fateful and destructive. After plague, famine, and sixteen years of war, Italy now experienced yet another disaster: the return of the Franks.

In the spring of 553, two brothers named Leutharis and Buccelin, seeking to take advantage of the chaos, led a force estimated at 75,000 across the Alps and into Italy. Both were leaders of the Allemanni, a tributary nation of the Merovingian Franks. (A portion of the confusion that seems unavoidable in reading scholarship supporting the existence of a historical German nation is this group, the linguistic progenitors of the German-in-French *Allemagne*.) As with numerous generals, both before and after, the two misjudged their opposition, "a puny little man, a eunuch of the bedchamber, used to a soft and sedentary existence."[29] Their evaluation must have seemed, at first, prescient, since throughout 553 and the winter of 554, their southward march through central Italy was largely unopposed, and they were able to acquire huge amounts of plunder "without haste."[30]

At Rome, where Narses had consolidated his own forces, the two brothers divided their armies, heading for still more plunder in Apulia and Campania. That satisfied Leutharis, whose ambitions ran more to wealth than power. He decided he had acquired enough to satisfy him for life, and headed north for home. There, he and his entire army encountered the demon, and "were decimated by a sudden outbreak of plague. . . ."

> their leader [Leutharis] became unhinged and began to rave like a madman . . . was seized with a violent ague and would tumble over backwards foaming at the mouth and with his eyes horribly distorted . . . the pestilence continued to rage until the entire army was wiped out.[31]

His brother was no luckier; when, evidently flattered into impulsiveness by his Goth allies, who persuaded him that a defeat of Narses would result in a kingship, he decided to force conclusions with the army he had been avoiding for more than a year. Near Capua, his bravado turned to recklessness; against another line of dismounted cavalry flanked with archers, the Allemanni formed up into a column before pressing the attack. It was a familiar maneuver that northerners had been using for centuries. Tacitus called it the *cuneus,* using the Latin word for wedge, since the column's head compressed into a point; the tactic was to throw axes, known as *francisca,* when within twenty to thirty yards of the enemy and then to charge home with sword and spear. Because the Allemanni, like the Franks, wore no armor at all above the waist and were rarely even helmeted, the results were predictably bloody. As the northerners advanced into a virtually undefended center of the imperial line, Narses ordered his divisions of archers to wheel left and right, where they had a clear line of fire. Though Agathias's account of the casualties—only five Franks survived, only eighty Romans killed—is unreliable, it communicates the scale of the victory, which Agathias compared to the Battle of Marathon.

Only months after receiving the news of the victory at Capua, on August 15, 554, Justinian issued a "Pragmatic Sanction"—a term of art in Roman law, covering statutes issued by the sovereign with a wide application and many components. Since the term refers only to the acts of an absolute ruler, many Pragmatic Sanctions dealt with matters of succession, but Justinian's most resembled one of those elephantine pieces of modern legislation with clauses on everything from agricultural price supports to excise taxes to research grants. The Pragmatic Sanction of 554 annulled all of Totila's actions (referring to him, apparently without irony, as the "tyrant"); all property sold under Ostrogoth rule became recoverable and slaves were restored to their former owners. The dole of free grain to citizens of Rome was restored and funds were allocated for repair of public buildings and aqueducts and for the salaries of municipal professors and physicians. The administration of the newly conquered territories was specified, with Sicily ruled directly from Constantinople, Corsica and Sardinia from Africa, and the remainder of Italy divided into eleven administrative districts. Most significantly, for the future of the Rome that was now "the Rome of the Popes," the governors of those districts were henceforth to be nominated by local bishops, and the lands of the Arian Gothic church were irrevocably granted to the Chalcedonian/Catholics.

One of the perversities of the empire's military success was the way it served to reinforce the political authority of the only competing institution in the Mediterranean world: the popes and bishops. For years, Justinian had shown his eagerness to coopt the power of the ecclesia in service of his dream of unification; in 546, he ordered a purge of pagan intellectuals—"illustrious and noble men, as well as a large group of grammarians, sophists, teachers, and doctors; after they were discovered and tortured they rushed to denounce one another; some were beaten, others flogged or put in prison . . ."[32]—as a matter of both policy and piety.

However, it was not the emperor's piety that was in question, but his assertion of doctrinal authority over the bishops. On May 5, 553, Justinian called for a council of bishops in order to reconcile yet another matter of Christological debate, this one about the incorruptibility of the Savior's body. At the Council of Constantinople (sometimes called the Fifth General Council) Justinian supported the position that the divine body had been changed in no respect "from the time of its actual formation in the womb."[33] His stance was essentially an endorsement of Monophysitism, and like Monophysitism it was a response to a religious paradox, this one about the resurrection. Evagrius described the bishops' reaction with even more passion than he had the plague ten years before: They would have nothing to do with skating even this close to heresy and argued that the body was corruptible, at least "in respect of the natural and blameless passions."[34]

As with all such debates, the ability of the post-Enlightenment mind to understand the passions with which such positions were attacked and defended is limited. No doubt that is one of the reasons that most descriptions of the debate focus on its political dimension, which was very real. The Fifth Council was a singularly important marker for the east-west fissure that would divide medieval Christendom between Roman Catholicism/Chalcedonianism, Eastern Orthodoxy, and a group of national/Monophysite churches in places like Armenia, Syria, and Egypt. Assuming that the dispute was *entirely* political, however, scants the importance of these matters of faith. It may have been not only in Justinian's self-interest to assert the ascendancy of emperor over pope, but that assertion also reflected his sincere religious faith.

———

Despite his command of the world's most powerful nation, and his equally powerful belief in the legitimacy of that nation's historical claims and the superiority of its national ideology, Justinian's campaign to make large portions of the world congenial to that ideology was problematic. Though the initial phases of the military operation were uniformly successful, and accomplished with surprisingly few troops, the occupied territories proved fractious. In Libya and in Italy, early triumphs were followed by a succession of mutinies and rebellions; after Belisarius's early victories, "Justinian had as yet no conception of the agonizing gap between military victory and final pacification. He was to learn of it, in Africa and elsewhere. . . ."[35]

The first lesson was administered in the form of a revolt by the Moors. No sooner was that revolt suppressed—by Belisarius's lieutenant, the eunuch Solomon—than the occupying troops mutinied. When the mutiny was ended (by the novel method of paying the mutineers for the time they had been fighting against their lawful commanders[36]) Libya became, "under [Solomon's] rule powerful as to its revenues and prosperous in other respects"[37] . . . so much so, in fact that modern roads, aqueducts, and fortresses throughout Libya and Tunisia are inscribed with Solomon's name.[38] However, Solomon's nephews, each appointed military ruler of a Libyan city, proved that talent for governance is not heritable. Procopius dismissed one of them, the ruler of Tripolitania (modern Tripoli), Sergius, as "exceedingly stupid and young both in character and in years, . . . the greatest braggart of all men."[39] Sergius so inflamed the only recently pacified

Moors that they revolted yet again, as did a succession of ambitious Roman soldiers and allies, including Solomon's erstwhile bodyguard, Gontharis. It was not until 548 that another of Justinian's seemingly inexhaustible store of gifted generals, John of Troglita, finally defeated the rebellion.

Similarly, between Belisarius's victory at Ravenna and Narses's at Taginae, the Italian peninsula suffered through ten long years of insurrection, famine, and—especially—plague. Hundreds of thousands were killed; Milan and Rome were depopulated. Markets were destroyed; agricultural productivity was devastated. In a dishearteningly familiar pattern, a succession of brilliant military successes was consistently followed by mutiny, rebellion, and one failure after another in the pacification of the newly conquered territory. There is a persuasive argument that Justinian, like the Roman generals indicted by Tacitus in the *Agricola*, "made a desert and called it peace."

However, an equally credible case can be made for the defense. Italy was no stranger to war before Justinian's reconquest began, nor was its future a peaceful one. But for the five years following the Pragmatic Sanction, the peninsula enjoyed a respite from war, as did Libya and Spain; Constantinople would rule southern Italy, generally peacefully, until the Normans conquered it in the eleventh century. In fact, anyone living in the years between the 554 and 558 might be forgiven for believing that Justinian's dream would be able to survive even the demon, since they would have had no way of knowing of its imminent return.

———

The *annus horribilis* that was 558 properly began in December of the preceding year when Constantinople was struck by a powerful and destructive earthquake, whose most notorious effect was the collapse of the original dome of the Hagia Sophia. No sooner had the rubble been cleared and reconstruction started than the plague returned.

The factors conspiring in a plague outbreak are related in such complex ways that predicting the rhythm of a pandemic is the sort of calculation that can frustrate a roomful of supercomputers. Estimating the moment when climate, flea behavior, food availability, and a dozen other variables will combine to cause a rat population crash is not impossible, but very nearly so. Reconstructing the various outbreaks of the pandemic from teasingly opaque evidence like funerary inscriptions, mass graves, and the rare eyewitness accounts, it seems that bubonic plague washed over the lands bordering the Mediterranean for nearly two hundred years, generally in waves between fifteen and twenty years apart,[40] but since no one believes the records to be exhaustive, this apparent oscillation might be nothing more than an artifact of inadequate evidence.

It seems certain, however, that no demon could strike a city the size of Constantinople without prompting someone to chronicle it. The best evidence that Procopius was either already dead, or living outside the capital, is that he failed to do so for the plague of 558 (nor did he record the destruction of the Hagia Sophia's dome in December of 557). We are fortunate that his self-proclaimed successor, Agathias, did.

> [The plague] now returned to Constantinople almost as though it had been cheated on the first occasion into a needlessly hasty departure. People died in great numbers as though seized by a violent and sudden attack of apoplexy. . . . According to the ancient oracles of the Egyptians and the leading astrologers of present-day Persia there occurs in the course of endless time a succession of lucky and unlucky cycles. These luminaries would have us believe that we are at present passing through the one of the most disastrous and inauspicious of such cycles.[41]

The most significant difference between this outbreak and its predecessor of 542 was that all the eyewitness accounts note that this plague had a powerful affinity for the young, who bore "the heaviest toll" according to both Agathias and John Malalas.

The reason is a simple one: Even a bacterium as formidable as *Y. pestis* can be frustrated by a prepared immune system. Like every pathogen, its life is a race to establish itself within a target organism before the organism's immune system can design killer T-cells to destroy it. Since that design-and-build process takes weeks, most plague infections are fought with generic weapons; and, most times, the fight is a losing one.

The exceptions occur when the bacterium encounters a preexisting blueprint, the acquired immunity known as a memory T-cell that is created after exposure to a pathogen.

Memory T-cells can be produced deliberately as well. Spurred by fears that *Y. pestis* may be used as a biological weapon, immunologists have experimented with dozens of possible vaccines. The earliest versions, typically weakened versions of the demon, required revaccination every three to six months, while some of the new ones, engineered rather than bred—one such vaccine is constructed from a sample of the F1 antigenic protein that appears on the demon's bacterial capsule,[42] another from the LcrV protein used by the Type III injectors[43]—seem to have far longer-lasting effects. Typically, however, no vaccine teaches the memory T- and B-cells as effectively as surviving the plague itself. In 558, a very high percentage of Constantinople's population older than sixteen had acquired that level of protection—while no one younger than that had any at all. The demon found better feeding among those born since its last visit, and it feasted on them from February until June.

By the summer of 558, as the demon's latest visit was starting to burn itself out, the "inauspicious cycles" brought yet another disaster to the imperial capital: Huns.

These particular Huns, the Kotrigurs, were close relatives of the peoples that Attila had used to terrorize the proto-European nations during the preceding century, but by the 550s their military capabilities had been diminished, not in battle, but by diplomacy.

Justinian was neither a great orator nor a skilled warrior, nor even a brilliant theologian (the last would probably have been his choice of profession). He was not physically brave or personally charismatic. He was, however, one of the greatest statesmen who ever lived, combining a grand vision for the empire he ruled with the ability of seeing a dozen moves ahead of his opponents. The emperor's complex—not to say Byzantine—diplomatic tactics make for dizzying reading. He had already mastered every tool of state policy short of war and had invented some new ones; he educated his enemy's princes, and sometimes found them wives. He gave them titles, money, and land. And he converted them to Christianity wherever possible, not merely out of his no-doubt real conviction that his God obliged him to do so, but because "baptism was a virtual acknowledgment of Roman overlordship."[44] With the Kotrigurs, he outdid himself, bribing one Hunnic tribe to attack another, informing the second that he would give them lands of their own as recompense for those taken by the first tribe, then paying off the first tribe as well once they objected to the original deal. It is correct to refer to this as buying peace, and peace was indeed what it bought, at least until 558. In the winter of that year, the Danube froze, and three Kotrigur armies crossed the defensive line turned entryway. Two entered Greece; one aimed for Constantinople itself.

The third army is remembered as much for its atrocities as its military success; even discounting the overheated recollections of later historians, this was a singularly violent invasion: nuns were raped, newborn babies trampled. But the terror tactics of the invaders were considerably less frightening to the emperor than was their strategic objective, for though less than ten thousand strong, the Kotrigurs were larger by far than any military force between themselves and Constantinople itself. Justinian had only two assets of any military value at all: The first were the palace guards—the Scholarians, a unit far more experienced at parade-ground spit-and-polish than the exigencies of real warfare against mounted Huns. The second was a retired fifty-three-year-old general.

Luckily, his name was Belisarius.

The general who had conquered Rome and Carthage for his emperor prepared to defend his city with fewer than five hundred trained soldiers. Filling out his ranks mostly with the peasants that had been driven into the dubious safety of the capital by the advancing Huns, Belisarius assembled his command at a village just outside the city's walls.

The general had regularly won more battles with shovels than swords, and was about to do so again; the troops spent their first days in the field digging a trench the entire circumference of the camp. He then disposed his missile artillery, two hundred slingers and javelin throwers, on either side of the Huns' route of advance. Finally,

Belisarius "armed" the peasants who made up the bulk of his force with wooden sticks—their only purpose: to make noise, and so convince the enemies that they were a larger force—and placed them behind his remaining cataphracts.

One can only wonder at the terror felt by an untrained peasant conscripted for a desperate battle against warriors that had been earning a reputation for ruthlessness for centuries, and which had only added to that reputation during their advance on the capital. Unlike the professionals that stood between them and the Huns, they had not learned to trust Belisarius's mastery of the battlefield. The archers and javelins had two purposes: to change the width of the Hun attack, and prevent a far larger foe from outflanking him; and to change the geometry of the mounted bowmen by making them ride knee to knee, unable to maintain a high rate of fire from bows. When four hundred Huns were killed before even closing on what appeared to them to be a sizeable force in a prepared position, they broke off the engagement. Justinian's gold looked far more attractive than his steel, so they accepted the inducements he had offered months before and retreated north to the Danube. Belisarius's skill had won one more battle.

As Justinian's shrewdness was to win one more war. For years he had been bribing not only the Kotrigurs to remain on their side of the Danube but also paying their mortal enemies, another Hunnic people called the Utigurs, as well. The emperor sent word to the Utigurs that the Kotrigurs were on their way, and that he would be most pleased if they were to relieve them of the gold they were carrying.[45] As the decade of the 550s wound down, Justinian appeared nearly as triumphant as he had at the beginning of the 540s.

The appearances, however, were deceiving. The structure that the emperor had built over the preceding twenty years was growing shakier every day, undermined both by the recurring threat of plague and the ambitions of the protonations that the plague helped bring into being. Narses could threaten Justinian's successors with a "thread they could not unravel" in the reconquered western territories partly because of another thread—a silken one—woven far to the east.

CHAPTER THIRTEEN

"This Country of Silk"

559–565

In the ninth *yanxi* year [166 C.E.] during the reign of Emperor Huan, the king of Da Qin [i.e., Rome] Andun, sent envoys from beyond the frontiers through Rinan [central Vietnam] to offer elephant tusks, rhinoceros horn, and turtle shell.[1]

—The *Hou Hanshou*

ONE OF ROME'S greatest defeats, as disastrous in its way as Adrianople and Cannae, was the defeat of Crassus by the Parthian army at the Battle of Carrhae in 53 B.C.E. The Roman prisoners taken by Parthians at Carrhae were later settled in a region of central Asia that, in the fullness of time, was temporarily conquered by a Chinese army who returned home with the legionaries, making them the first, though sadly undocumented, contact between Rome and China. More than two hundred years later, a commercial mission departed Rome's Mediterranean empire on a world-girdling voyage east. In 160 C.E., the overland route was still occupied by the Parthians, whose hospitality toward Roman ambassadors was uncertain, so to avoid them, the embassy took the sea route. As best as can be reconstructed, the expedition crossed the Mediterranean to Alexandria, traveled south to the Axumite Kingdom in what is now Ethiopia, there taking ship across the Indian Ocean and the Bay of Bengal, through the Strait of Malacca, and then north to a landing on the coast of Vietnam near what is today Hanoi. From there they traveled overland six hundred miles northeast to the most populous city in all of Asia, Luoyang (alternatively Lo-yang), the court of the Han in Hunan province.

Like Rome, China had ruled an empire for centuries before she was ruled by an emperor. The first of the many dynasties into which Chinese history is conventionally divided is the Shang, whose murky reign begins sometime around 1600 B.C.E. Far better documented are their successors, the Chou, originally a tributary nation of the Shang from present-day Shaanxi province, who deposed their onetime masters in a series of battles from 1122–1111 B.C.E. and ruled until 255 B.C.E. During their last few centuries—another entry in the history books conventionally surrounded by quotation marks, the period of "warring states"—China was as culturally rich as Athens in the roughly contemporary Periclean age, producing both the military thinker Sun Tzu and the philosopher Kung fu-Tze, more widely known in the west under his Latinate name: Confucius. But not until Shih Huang-ti founded the Qin Empire in 221 B.C.E. did China have her first true emperor. In the Roman style, Shih left no successor, and his empire did not long survive his death. The same cannot be said for the imperial dynasty that followed. From 202 B.C.E. to 220 C.E., the Hans ruled what had been established by Shih as the traditional "China," from Mongolia in the north to the Yangtze in the south.

Edward Gibbon claimed that the era that saw the greatest combination of human happiness and prosperity occurred between the death of Domitian and the death of Marcus Aurelius. Had he been writing about second-century Asia instead of the Mediterranean, he might have said the same for the Hans, who oversaw not only the world's most extensive irrigation economy and the invention of water clocks, paper, and block printing but also an agricultural surplus large enough to foster an enormous population explosion. By the time of the Imperial Census of 2 C.E., Han China was home to nearly 60 million, probably about the same size as the Roman Empire of the same date.[2] (Together, the two empires included nearly half the entire world's population.)

Trade between the east and west, broadly speaking, was more than seven hundred years old by then,[3] but commerce authorized by the two great empires dates only from the time of Augustus, who had established a trade mission in southern India near what is now Pondicherry, and almost certainly handled Chinese merchandise as well as Indian.[4] During the reign of Claudius, a fortunate trader named Annius Plocamus was blown by the monsoon across the Indian Ocean to Sri Lanka, and by the end of the first century, Rome had established direct trade along a sea route that led from Sri Lanka to China itself.[5]

The huge distances and costs required by such transactions could only be justified by cargos that were extraordinarily valuable, produced in a location with an insuperable comparative advantage. Such would be the case with Ceylonese cinnamon in the seventeenth century, Peruvian silver in the sixteenth, and Persian Gulf oil in the twentieth. But while precious stones and spices offered some temptations to Rome's merchants—recall that the protection payment made in 409 to Alaric included three thousand pounds of pepper—neither was sufficient to warrant the sort of expense needed for the years-long expeditions to China. A clue to the commodity that dominated "international" commerce in late antiquity is found in Ptolemy's *Geography*, which named the lands ruled by China *Sericai*. The people living there were still called, in the time of Justinian, the *Seres*, an adaptation of the Chinese name for the mulberry leaf-eating moth with the Linnaean name *Bombyx mori*. In its larval stage, the moth is more commonly known as the silkworm.

Like all Lepidoptera, the insect family to which moths and butterflies belong, *B. mori* is spectacularly skilled at changing itself: from egg to larva to chrysalis to moth. Only one of those stages, however, can persuasively be said to have changed history. *B. mori*'s version of the cocoon needed to morph from legged caterpillar to the "scale-wings" that give Lepidoptera their name combines two liquid proteins into a single filament—so fine that a single cocoon can produce a thread five hundred meters in length weighing less than a gram—that hardens when exposed to air.[6] In a legend first set down by Confucius, the *B. mori*'s secret was discovered by the accidental fall of a silkworm cocoon into a cup of tea. The teacup in question was in the hand of a

Chinese princess of the Chou Dynasty, Xi Ling-Shi, who was so en-
chanted by the silken thread formed when cocoon met hot liquid that
she became the patron of China's silk industry—the "Lady of Silk."
Legend or no, during the centuries separating Xi Ling-Shi from the
arrival of the Roman ambassadors, Chinese farmers learned how to
cultivate the silkworm by feeding it the leaves of the white mulberry
tree, then steaming or boiling the cocoons before the chrysalis ma-
tured, and pulling—the term of art is "reeling"—the filaments into
useable strands of silk.

For centuries, China monopolized not only silk manufacture, but
consumption. Silk did not become a significant item of trade with the
west until the first century B.C.E., simultaneous with, and possibly as a
result of, the increased political importance of steppe nomads.[7] As the
great nomadic confederation known as the Hsiung-nu [or Xiongnu]
became increasingly successful in raiding more settled Chinese territories to the east, the Chinese empire became amenable to the idea of
paying them to stop. More payments made the nomads wealthy,
which whetted their appetite for the luxuries of civilized life. By 50
C.E., that appetite had become large enough that the annual "gift"
from the ruling Han dynasty included ten thousand rolls of silk, and
the west of China soon became saturated with the stuff[8]—a historical
irony, should they be proven to be, as frequently claimed, the ancestors of the famously unpacifiable Huns.

The oversupply of silk on the western side of the Great Wall eventually proved a powerful inducement for overland trade, along the
land route long established for caravans carrying exotica like African
ivory, jewels, and spices: from the Great Wall across Chinese Turkestan
to the Tarim desert, which was skirted either to the north, through
Turfan, or the south, through Khotan, then to Samarkand, Merv,
northern Iran, Ctesiphon, up the Euphrates and thence to Antioch
and the Mediterranean.[9] By the first century C.E., silk was by far the
most important merchandise making the journey. Prominent Romans
lurched from deploring the luxurious fabric—Seneca wrote, "Wretched
flocks of maids labor so that the adultress may be visible through her
thin dress, so that her husband has no more acquaintance than any . . .

foreigner with his wife's body"[10]—to gorging on it. Caesar owned a silk awning, and dressed Cleopatra in a see-through silken garment. Vespasian might be remembered as the archetype of the austere Roman aristocrat, but still wore silk when celebrating his triumph.[11]

For the Roman world, silk was not merely a fabric, and not even just a valuable commodity, like gold. "It possessed special significance. It was the attire of the emperor and the aristocracy, an indispensable symbol of political authority, and a prime requirement for ecclesiastical ceremonies."[12] So important was the precious stuff that it required special legal attention; laws regulated the use of silk by dyers of purple cloth, in order to maintain control of what was an aristocratic and imperial privilege. Even before Justinian's recodification of Roman law, it already criminalized the wearing of gold-embroidered silk cloaks, any silken item dyed the best quality of purple, and even any imitation of the high-quality purple dye.[13]

Despite their appetite for the extraordinary fabric—lighter than linen, stronger than wool, practically transparent (Lucan's word is *perlucent*)—it was not until the late second century that Rome's people learned the source of the fabric. Some of their confusion is a result of the small quantity of wild silk—cocoons gathered when already chewed through, resulting in dozens of short filaments, rather than the single long one reeled from one that had been harvested while the pupa was still alive—already known in the Mediterranean. Pliny, in his *Natural History*, had some bizarre notions about wild silk cultivation (including the idea that the silk moths "grow shaggy hair and equip themselves with thick coats to combat winter, scraping together down from the leaves with their rough feet. They compact this into fleeces, card it with their claws and draw it out into the woof, thinned out as if by a comb, and then they wrap this round their body."),[14] but at least knew that it was made by silkworms turning into moths. He was, however, convinced that the Chinese fabric was a different substance altogether, one harvested from trees: "The *Seres* [i.e., Chinese] are famous for the woolen substance obtained from their forests; after a soaking in water they comb off the white down of the leaves."[15]

Despite, or perhaps because of, the mystery of its origin, by the time Constantinople was the imperial capital, silk was the most important trade good crossing the empire's borders. As such, payment for it represented a huge outflow of gold, though one perverse oddity of the silk trade is that its entry into the empire was sometimes paid for by . . . silk. The heavy Chinese silk was frequently unwoven by Syrian weavers, who then turned it into a gauzy material re-exported to China. In one of the longest-lasting hustles in commercial history, the Parthian middlemen of the first century managed to convince their Chinese customers that Romans actually possessed silkworms of higher quality than did the Chinese, a myth that they used to keep the prices from skyrocketing.[16]

However, by the early fifth century, the Chinese were very well aware of the existence and value of their monopoly. They had also learned that they were playing catch-up with their own middlemen, who worked feverishly to protect their status: A fifth-century Chinese history states "The kings of Ta Ts'in [the Chinese name for the Romans] always desired to send embassies to China, but the Parthians [NOTE: He almost certainly meant Sassanians, as this was written in the fifth century] wished to carry on trade with Ta Ts'in in Chinese silks and therefore cut them off from contact."[17] Cosmas Indicopleustes, a sixth-century Greek sailor and traveler to India, wrote a rather peculiar book (he argues against the consensus belief in a spherical earth by rewriting biblical passages in order to prove, like some sixth-century Immanuel Velikovsky, that the rounded heavens sit above a flat earth) entitled *Christian Topography*, in which the following passage appears:

> Now this country of silk is situated in the remotest of all the Indies, and lies to the left of those who enter the Indian sea, far beyond the Persian Gulf and the island [NOTE: Sri Lanka] called by the Indians Selediba and by the Greeks Taprobane. It is called Tzinitza, and is surrounded on the left by the ocean, just as Barbaria [NOTE: Somalia] is surrounded by it on the right. . . . The country in question deflects considerably to the left, so that the loads of silk passing by land through

one nation after another reach Persia in a comparatively short time, while the route by sea to Persia is vastly greater . . . That is why there is always to be found a great quantity of silk in Persia.[18]

Thus, while the reconquest of the west was Justinian's highest priority, scarcely a month passed during his long reign where he did not attend to his eastern frontier as well, where his strategic objective was not territory, but trade. As early as 525, the emperor was trying to use Ethiopians as proxy buyers of silk from the Indians, for they "might make themselves much money, while causing the Romans to profit in only one way,"[19] that is, by keeping Roman money out of Persian pockets.

And it was, by any measure, a *lot* of money. By statute, raw silk was purchased from the Persians at fifteen *solidi* to the pound.[20] Since one solidus weighed about 70 grains, a pound of silk cost nearly two and a half ounces of gold; a hundredweight, 240 ounces, or 20 Roman pounds (15 modern pounds). As a consequence, for every ton of silk traveling west from China to Rome, four hundred pounds of gold made the trip the other way, a significant proportion ending up in the hands of the Sassanid middlemen. No reliable records document the total annual trade, but a single ox wagon (a *cursus publicus*) of late antiquity could carry up to 1,500 pounds,[21] and dozens of caravans headed through Mesopotamia on the Silk Road every year. If those caravans carried forty tons or so of silk (a modest enough estimate) west, then up to eight tons of gold—more than 7 percent of the entire imperial treasury at the time Justinian took office—headed east every year.

The silk trade represented both a giant windfall for the Sassanid Persians, and an equally huge drain on the empire; much of Justinian's and Khusro's maneuvering makes sense only in light of their respective desires to break or preserve the status quo. These maneuverings stretched as far north as the shores of the Black Sea—Richard Frye, the great historian of ancient Iran, suggests that the wars in Armenia and Lazica were "a prelude to the ambitious dreams of Khusro to control the trade of the silk route to China"²²—and as far south as Ethiopia. But Persia's most vulnerable economic flank—and, therefore, Justinian's lifelong strategic objective—was the Arabian peninsula, a land that, by the rules of sixth-century realpolitik, was important to the two empires precisely to the degree that each could deny it to the other.

———————

Contemporary visitors to Khusro's court at Ctesiphon write that the palace contained three empty thrones: one for the ruler of the nomadic horsemen of the Central Asian steppe (the Great Khagan); one for the emperor of China; and one for the Roman emperor. The thrones were there in case any of these sovereigns came, as supplicants, to the Persian king of kings. Their presence indicated not merely the arrogance of the Persians but also their blinders, for the three compass directions represented by the thrones—north, east, and west—failed to look south, toward the Arabian peninsula, where the last act of the Sassanids would be written.

That portion of Arabia lying south of the Syrian desert was described by Procopius as "a great tract of land [where] absolutely nothing grows except palm trees . . . a land completely destitute of human habitation."²³ This is not precisely accurate; cities like Mecca had grown up around caravan oases, and the south Arabian coast— biblical Sheba—had been settled for millennia. Yemen, in fact, was the most powerful local "state" in Arabia until the rise of Islam, with an intensive irrigation system fed by the remarkable Marib Dam, a huge structure more than fifty feet high, whose sluice gates poured the rainfall captured from the Indian Ocean monsoons into hundreds of miles of canals. Yemen was so attractive, in fact, that Augustus himself had attempted its conquest in 25 B.C.E.

In 404, the historian Synesius praised the bravery of one group of Arab soldiers, describing them as "by nature true descendants of Homer."²⁴ The Homerites, or Himyarites, in one of late antiquity's stranger stories, were neither pagans nor Christians at the time, but Jews, and had been so since the late fourth century when a local chieftain brought two rabbis back from the city of Medina. By the middle of the sixth century, the Jewish Arab Himyars were ruling Yemen, which prompted one of the early proxy wars between Constantinople and Persia. In 517, the Himyar king Yusuf As'ar Yath'ar (known in Arab tradition as Dhu Nuwas) massacred hundreds of Christians, mostly from Ethiopia, or, as it was then known, Axum. The Axumite king, Ela Atsheba, responded with a punitive expedition, which was defeated by Yusuf; but since Yusuf had asked for aid from the Persians in resisting the Christian Axumites, the Axumites went to Justin, who provided ships and supplies for a new invasion. In 525, the Axumites entered Yemen a second time, this time defeating and killing Yusuf and replacing him with an Axumite general.[25] Procopius describes an Axumite embassy's attempts to make common cause with Constantinople against Persia, but by then the empire had chosen a new proxy: the Ghassanids.

The fifteen-year war between Roman-supported Ghassanids and Persian-supported Lakhmids had many causes,* but one was surely the desire of Justinian to circumvent Khusro's silk tariff; in the words of modern historian Robert Browning, "control of the [Red Sea] straits . . . enabled [the empire] to circumvent a possible Persian blockade. . . ."[26] Even had Justinian succeeded, however, no trade route to China could ever be completely secure; for every move in this sixth-century version of the Great Game that would one day obsess the nineteenth-century empires of Russia and Great Britain, a counter-move existed. So long as China was the sole source of silk, Arabia remained a uniquely valuable bit of desert.

So long as China was the sole source.

The oasis known as Ho-t'ien, Hotan, or Khotan, is one of the few islands of permanent habitation in the far western Takla Makan desert that traditionally separates China from Central Asia. A stop on the Silk Road, Khotan also marked the rather vague border between

---

* It also had some unanticipated consequences, including training a large number of Arab soldiers in modern military tactics, a favor they would repay early in the following century.

Indian and Chinese civilization, one of the places through which Buddhism made its way from India into China. Khotan was one of civilization's tidal pools: those places (like Nisibis) where several civilizations came fruitfully in contact.

Fruitfully; not peacefully. Khotan would be fought over by Chinese, Mongols, Tibetans, and even Moghuls. The Chinese retreated through the oasis when defeated by the Arabs at the Battle of the Talas River in 752. Even today, it is a hotbed of Muslim/Uighur agitation against the Chinese.

The early twentieth-century historian Sir Aurel Stein translated the legend of the introduction of silk to Khotan in the first century[27]— a legend that establishes the secretive character of the secret of silk cultivation:

> [T]he ruler of China was determined not to let others share their possession, and he had strictly prohibited seeds of the mulberry tree or silkworm eggs being carried outside his frontiers.[28]

So, when a Chinese princess fell in love with and married a prince of Khotan (probably an Aryan, closer ethnically to his Indian ancestors than his Chinese in-laws) he reminded her that if she wanted to continue wearing silk robes, she would be well advised to include mulberry seeds and silkworms in her trousseau.[29]

Four hundred years later, history repeated itself. Sometime between 551 and 553, two monks, probably Nestorians, arrived at Justinian's court, having traveled from their home in Khotan, well aware of the emperor's desire to find some way to avoid enriching Persia. The monks

> had learned accurately by what means it was possible for silk to be produced in the land of the Romans. Whereupon the emperor made very diligent enquiries and asked them many questions to see whether their statements were true, and the monks explained to him that certain worms are the manufacturers of silk, nature being their teacher and compelling them to work continually. And while it was impossible to convey the worms [to Constantinople] alive, it was still practicable and

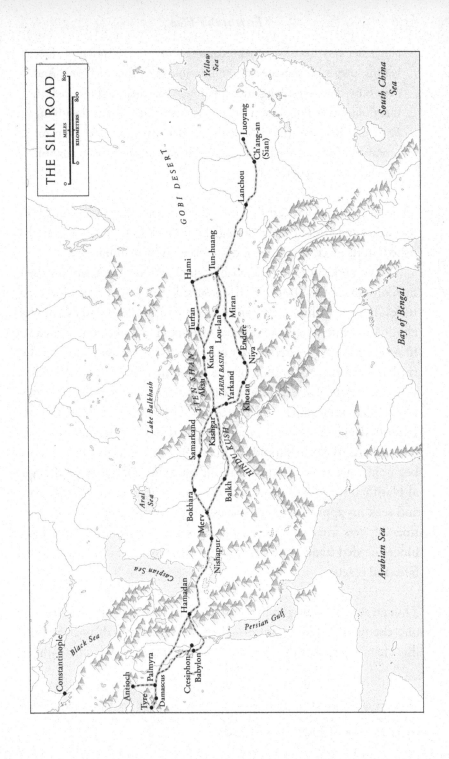

altogether easy to convey their offspring. Now the offspring of these worms, they said, consisted of innumerable eggs from each one. And men bury these eggs, long after the time when they are produced, in dung, and after thus heating them for a sufficient time, they bring forth the living creatures. After they had thus spoken, the emperor promised to reward them with large gifts and urged them to confirm their account in action.[30]

Their second journey, carrying the eggs—in legend anyway—in a hollow cane, ended the Persian stranglehold on the silk trade, and established, in its place, an imperial monopoly on its manufacture.

The consequences were probably more serious in Ctesiphon than in Constantinople. Trade in Chinese silk would continue for centuries, but the ability of Persian middlemen to collect a premium was eliminated, with an annual net loss of Roman gold likely measured in tons. Even so, the greater impact may have been in Arabia. In 554, the rivalry between al-Mundhir of the Lakhmids and al-Harith ended with the death of the Lakhmid king and the complete defeat of his army, a day traditionally celebrated as "The Day of Halima" (for al-Harith's daughter, Halima, who "perfumed the [Ghassanid] champions and clad them in shrouds of linen"[31]). The battle also marked the beginning of the retreat of Roman influence in Arabia.[32] Even though al-Harith traveled to Constantinople in 563 to meet with Justinian and seek his approval for his succession plans, the emperor was considerably less interested in subsidizing a proxy army to defend a no longer needed trade route.* Not after twenty years of diplomacy, warfare, and plague.

————

Though Arabs residing in Syria experienced the same plague as everyone else in the Mediterranean—the sixth-century Arab poet Hassan ibn Thabit, describes the plague as "the stinging of the *djinn*"[33]—there

————

* Memory of the subordinate relationship between al-Harith and his imperial sponsors has proved resilient; in October 2004, Arab rulers cooperating with the west were accused, by Osama bin-Laden, of being "the new Ghassanids."

is hardly any record of the demon in Arabia. It is tempting to argue that the isolation of central Arabia offered a barrier to the disease that was killing millions to the north of the desert border. A factor as obvious as the desert's daytime heat may be significant; as we have seen, the delicate balance that pushed bacterium, flea, and rat out of their rough equipoise occurred only within a narrow temperature range, and only when that balance was upset did *Y. pestis* jump into other mammalian populations. Equally persuasive is the idea that the relative sluggishness of desert trade—caravans that were slower than both ships and the ox carts that traveled the empire's roads—made for a poor highway over which to carry an infection that burned itself out in a week or so. More important than any climatic or topographical barricade, however, was the arithmetic of the basic reproductive number, since the most sensitive variable in epidemiology is *c*, or the rate of contacts. Arabia's relatively low density of population—even in the most urbanized part of the peninsula (Yemen) cities were only a fraction the size of Alexandria or Antioch—was a firebreak more imposing than a hundred miles of uninhabited desert.

Absence of evidence is not evidence of absence. Nonetheless, whatever the incidence of bubonic plague in the territory that Pliny had named *Arabia Felix*, it is certain that the Arab tribes that survived the sixth century were an enormously resilient people. The Romans and Persians had recruited them for their ferocity, trained them in tactics, and disciplined them through decades of war. They entered the seventh century prepared for an empire of their own, a people "dangerous to provoke and fruitless to attack."[34]

The relative vigor of the Arabs also made them the final milestone along the road to premodern Europe. The fragmented territories conquered by Scipio, Caesar, and Trajan were united by Roman law; once the writ of that law could no longer be enforced, their coalescence into discrete kingdoms may have been inevitable (though, as we have seen, predicting which kingdoms—Franks over Goths, for example—is fundamentally impossible). Giving those separate kingdoms a common identity in the absence of a shared earthly sovereign, however, required the presence of a common enemy.

It is scarcely original to observe that the Arabs and their defeated-and-converted conquests, by giving Europe such an enemy, defined it;* this is one of the central arguments of Henri Pirenne's great work *Mohammed and Charlemagne*. Europe—or Christendom, as it would more popularly be called for centuries—is at least partly a consequence of the Arabs who succeeded al-Harith and al-Mundhir. It is also worth noting that while the Arab armies of the seventh century were no better protected from plague than the Illyrians of the sixth and suffered through attacks at Ctesiphon, Jerusalem, and Basra, in which hundreds of thousands perished, the Arabian desert itself remained relatively free of the demon, which gave its children something few lands of late antiquity could boast: a plague-free home.

Another virtually plague-free home was found in the only other empire on the Eurasian land mass to rival Rome or Persia. China's escape from the demon's worst ravages may even be the most important difference in the otherwise parallel histories of the Roman and Chinese empires. Both defeated and absorbed their predecessors—Greece, Egypt, and Carthage for Rome; Yan, Qi, Wei, Zhao, and Chu for China. Both faced the same early epidemics of smallpox and measles, the same political instability, the same stormy third century. At the dawning of the sixth century, the Chinese empire, like Justinian's, faced a similar loss of traditional land to barbarians (the so-called Sixteen Kingdoms of northern China) and reconstituted itself in an eerily similar way, led by an emperor of enormous talent and peasant birth named Yang Chien.†

Only afterward, "historical developments in these two regions began to diverge permanently."³⁵ The reasons were, frankly, demonic.

---

* And not always as an enemy; long before the Arab conquests brought them to the borders of Europe, a tribe in northern Arabia created *Udrn*, formal love poetry that combined chivalric romance with Christian chastity and would be the direct ancestor of the *amour courtois* of the Middle Ages. (Irfan Shahid, *Byzantium and the Arabs in the Fifth Century*, Washington D.C., Dumbarton Oaks Reseach Library, 1989.)

† Yang Chien, better remembered as Wen-ti, and the founder of China's Sui dynasty, also had as his partner a wife that appears to be both as formidable and as much a magnet for scandal as Theodora.

Though caravans traveled from western China to the eastern Mediterranean for centuries, they carried silk, rather than the grain that would keep rats alive for the two-hundred-day-long journey. Consequently, the pandemic that lasted for nearly two hundred years in the Mediterranean was confined, in China, to two cities: Chang'an and Loyang.

———

As the decade of the 560s opened, Justinian had been emperor-in-fact for more than forty years. The empire he ruled was both far larger and more fragile than the one he inherited; imperial dominion extended from the Guadalquivir River to the Euphrates; but, only two years earlier, invaders had camped outside the walls of the capital itself. Taxes were being collected from Italy and Libya, but the armies needed to defend those territories cost even more, turning the imperial administration into a machine barely able to feed itself. Justinian had successfully enforced the authority of the emperor over religious life, but the conflict between Monophysite and Chalcedonian/Catholic was as unresolved as ever, to say nothing of the tension between the Bishop of Rome and the emperor and other bishops alike. Most seriously, the agricultural-economic base of the empire was so severely depleted by plague-caused population decline—twenty-five million fewer people lived in the empire at the end of the sixth century as did at its beginning—that the army it supported could field less than a third the strength it could only decades before.

While the army's capabilities were diminished, its responsibilities were not; in 555, the steppe people known to history as the Avars revolted and were massacred. Twenty thousand survivors fled west along the Scythian highway, reaching the Roman garrison north of the Black Sea in 557. They demanded settlement lands and were sent to meet with Justinian in Constantinople, where he accepted them as allies, but refused to grant them land.[36] Conscious of the danger posed by this latest incarnation of the nomadic horsemen who had been threatening the empire since the Huns, Justinian scrambled to put sufficient troops in the field to police his new *federates*, and attended to reconstruction of the long walls that protected the isthmus on which the city of Constantinople was built, giving himself the equivalent of a triumphal procession and oration, despite the fact that his "victory" was over a wall, and he returned with no captives, and no treasure.[37]

In 561, the Avar king, the Khagan Bayan, again demanded land in Thrace, south of their north-of-Danube temporary settlements, and was again denied. Frustrated by Justinian's intransigence and stymied by his fortifications, Bayan followed in the bootsteps of the Huns, Goths, and Vandals who had previously migrated west, evidently intending to get their land from the Franks. If that was his intent, he was to be thwarted again. Defeated by the Merovingian army, the Avars returned to the lower Danube, where they waited (until Justinian's death) to found a brief but enormous empire that at one point stretched across the entire central European plain, all the way to the Black Sea—incidentally, becoming one of the reasons that the Lombards were willing to accept Narses's invitation to march into Italy.[38]

In 565, as if to remind the empire that the power of its greatest enemy was undiminished, the demon renewed its assault on the Mediterranean. As recorded by an eyewitness, Paul the Deacon, "a very great pestilence broke out, particularly in the province of Liguria,"[39] though it did not remain there, rapidly spreading as far north as Bavaria. This was, by then, the third decade of the plague years, and the landscape of deserted homes, abandoned children, and unburied bodies must have started to become wearily familiar. Nonetheless, Paul's descriptions remain horrific: The world, he wrote, was:

> brought back to its ancient silence: no voice in the field, no whistling
> of shepherds . . . the crops outliving the time of the harvest, awaiting
> the reaper untouched, pastoral places turned into a sepulcher for men,
> and human habitations into places of refuge for wild beasts.[40]

This latest assault on his empire might have been expected to be yet another source of great anxiety to Justinian, now more than eighty years old. He was, by all accounts, spending ever more time indulging his love of doctrinal subtlety, Christological controversies, and the details of the church's calendar and hierarchy. The man who had caused the entire corpus of Roman law to be rewritten spent his last years on such minutiae as the dates of Epiphany and the proper punishments for breaches of clerical discipline. In October 563, for reasons that seemed obscure even to his contemporaries, the old emperor made a journey of pilgrimage to the Church of St. Michael in Anatolia, probably the first time in decades he had been more than a few days from the capital.[41]

It was also to be the last. Unaware of the latest ravages of *Y. pestis* on the land he had returned to the empire, on November 14, 565, without any warning save his advanced age, Justinian died.

The emperor may have outlived his usefulness; he certainly outlived everyone else, including Theodora, Germanus, John, Anthemius, Tribonian, Procopius, and even Belisarius, who predeceased his emperor by eight months. Of Justinian's best and brightest, only Narses, still ruling Italy in the emperor's name, remained alive to experience the rather unfortunate rule of Justinian's successor, whose accession was even more suspicious than Justinian's own. Having survived so much in his long life—not least his own visit, twenty-three years previously, from *Y. pestis*—Justinian was, even more than most autocrats, convinced of his immortality. As a result, he had failed to name a successor (the most obvious choice, Germanus, had been dead for fifteen years) until the night of his own death, and possibly not even then: The only witness to the emperor's supposed choice was the Praepositus of the Sacred Bedchamber, Callinicus, who brought the news to Justinian's nephew Justin[42] that he had succeeded to the imperial throne.

Thirty-eight years after Justinian and Theodora had been crowned in the old Theodosian Church, Rome's last great emperor was laid in state on a bier covered with gold leaf. A day later, the body was carried from the palace followed by Justin and his wife, Sophia; the city's patriarch, senators, and other high officials; generals, soldiers, and members of the Excubitores, through streets lined with a hundred thousand weeping citizens of Constantinople, to the Church of the Holy Apostles, where Justinian was interred in a porphyry sarcophagus next to his "most pious consort," Theodora. He was wrapped in a shroud that already contained, like a prewritten obituary, embroidered depictions of Justinian's achievements. The shroud was made of silk.

———

The thirteen-year-long reign of the mad Justin II—at one point, bars were affixed to his windows to frustrate any suicide attempts[43]—can count among its achievements the unchecked invasion of the Lombards into Italy, the establishment of the Avars on the Danube, and the encroachments of the Slavs into the Balkans and Greece. The empress, Sophia, prevailed on her husband to choose a general, Tiberius, as his successor in 578, but while sane, Tiberius was no better able to preserve Justinian's achievements than his predecessor. He was, in turn, replaced by his most successful general, Maurice, who was able, for twenty years, to stem the flood of bad news, partly because his opposite number in Persia confronted even worse problems. So bad, in fact, that in 591, facing an internal revolt, Khusro II sought an alliance with Maurice as the best of a set of bad alternatives, granting him Armenia and the Mesopotamian frontier cities in return for aid.

Maurice's successes were circumstantial and temporary. In what is probably the most familiar trope in the history of the Roman Empire, Maurice failed to ensure the loyalty of his own army. Though a gifted diplomat and strategist—he wrote the definitive text on warfare, the *Strategikon*, and organized the western provinces under governor-generals, or *exarchs*—the emperor apparently assumed that authority *ex officio* could replace personal command, which he left to carelessly chosen underlings. He thus forgot both the lessons of Diocletian, who never let an army stray too far from a tetrarch, and Justinian, whose talent for surrounding himself with loyal retainers was legendary even in his lifetime. The results were predictable. In 602, the imperial troops on the Danube revolted, chose a middle-grade officer named Phocas to lead them, and marched on Constantinople. Allied with the revived Blues and Greens, Phocas pursued Maurice and his family to Nicomedia, where they had fled, and executed the emperor's children one by one while their father was forced to watch. Only after Maurice himself was finally beheaded did Phocas declare himself the heir of Augustus and Constantine.

Or, more properly, Caligula and Nero. Not since the worst excesses of the early empire was it ruled by a man with such an appetite for cruelty and torture. Phocas's brutality was not only pathological, it was suicidal; at the very moment that he was under attack in the east from Khusro II, Phocas recalled his best general from Mesopotamia, apparently for the pleasure of burning him alive.[44]

Phocas was not, of course, the only curse afflicting the empire. For the bulk of the population, in fact, he was probably the least of them: Another outbreak of plague hit Rome in 590, killing Pope Pelagius; at the installation of his successor, Pope Gregory the Great, the new pontiff demanded that the residents of the Holy City join him in a litany, at which eighty people fell dead.[45] The demon traveled to Ravenna in 591, and to Antioch in 592 (when the daughter and grandson of Evagrius Scholasticus died). In 597, St. Demetrios of Thessalonica wrote of the "all-consuming and all-destroying plague" arriving in his city.[46] Elias of Nisibis reported another outbreak in Syria in 599.

To the average citizen, it must truly have seemed that God had forsaken the empire. It must, therefore, have seemed like deliverance when Heraclius arrived.

The son of the exarch of Carthage, Heraclius was picked, by his father and uncle, to lead a combined naval and military assault on Phocas, which he concluded successfully in 610, becoming emperor in his place. The crown was by then more a burden than a prize, what with the Avars overrunning the Balkans and the Persians threatening to revive the dreams of not only Khusro Anushirvan, but Xerxes: In 611, Khusro II's armies took Antioch; Damascus in 613; and, in 614, Jerusalem, seizing as spoils the artifacts of the crucifixion from the Church of the Holy Sepulchre.

For twelve years, Heraclius demonstrated a remarkable equanimity in the face of these setbacks, using the time to consolidate his remaining territories, rebuild his army and—most important—to formalize a financial alliance with the church. On Easter Monday, 622, the emperor led his new army into battle. Over the course of the next five years, Heraclius conquered one Persian army after another, trusting the walls of Theodosius and Constantine to preserve the capital from attacks by both Sassanids and Avars. In 627, he took Ctesiphon itself, and on September 14, 628, returned to Constantinople with the pieces of the True Cross that Khusro II had removed from Jerusalem fourteen years before.

The campaign against Persia was to be the high point of Heraclius's reign. While he was writing the last chapter in the thousand-year-long war that had defined the classical world, another age was beginning. Nearly sixty years before, five years after the death of Justinian, a son had been born to the Hashim, a prosperous family of merchants— trading in, among other things, Chinese silk—living in the city of Mecca. His name was Abu al-Qasim Muhammad ibn Abd Allah ibn Abd al Muttalib ibn Hashim: Muhammad, the Messenger of God.

Yarmuk

636

THE ROCKY PLATEAU known as the Golan Heights forms a V, criss-crossed by ravines, that runs downhill from the base of 9,000-foot-high Mount Hermon in the north to the Jordan valley in the south. On the plateau's east side is the Damascus plain; on its west, a 2,000-foot-high ridge separates the Heights from the Sea of Galilee. At the vertex is the tangled, ravine-cut valley of Yarmuk, named for the eponymous river that formed it, flowing east to its confluence with the Jordan. On June 9 and 10, 1967, the Yarmuk valley was the site of the final assault of the Six-Day War, when Israeli troops under colonels Aharon Avnon and Mordechai Gur[1] attacked the Syrian army in the north. Avnon and Gur fought over ground soaked with the blood of centuries—thirteen centuries, to be precise—more than a thousand years of conflict between the armies of the world's youngest monotheistic religion and those of its predecessors.

Ceaseless conflict, in fact: Israeli against Arab, Briton against Ottoman, Crusader against Saracen. Geography alone—two martial societies occupying the same real estate—made conflict inevitable whenever an Arab nation asserted itself north of its desert home; as early as 273, the desert kingdom of Palmyra fought the empire to a standstill for more than a year. The war that began with the first battle of Yarmuk, however, was different, a fundamental religious conflict resolvable only through violence, and sometimes not even then.

———

The revelations received by the forty-year-old Muhammad—later collated into the Qur'an—began in 610. Three years later, he began his public ministry; two years after that, his implicit criticism of the merchants of Mecca provoked their opposition, which culminated in the *Hegira,* or emigration to Medina, in 622, the first year in the calendar of Islam.* In 630, having established his political and religious authority in Medina, and defeated his onetime Meccan opponent, Abu Jahl, at the Battle of Badr, Muhammad entered Mecca in triumph.

Two years later, having united Arabia, he died in Medina. The internal conflicts that followed his death were resolved in favor of the dynastic marriages that Muhammad had pursued in service of a consolidated Islamic nation. As a result, the Prophet's first successor was Abu Bakr, the father of one of Muhammad's wives, Ayesha; his second—the second caliph—was 'Umar, the father of Hafsah, another of Muhammad's wives. One of Muhammad's daughters, Um Kulthum, was in turn married to Uthman, who would be the third caliph, and another, Fatima—by all accounts his favorite—was married to the fourth caliph, Ali.

Under 'Umar, the Muslim armies looked northward, where the barricades built to contain them were ready to fall. Just as the Ghassanid Arabs had been weakened by the loss of Roman support after the end of the silk monopoly, the Lakhmid Arabs had suffered at the hands of their sponsors: In 600, Khusro II, reinstalled on his throne by Maurice, had imprisoned and executed the rebellious Lakhmid king, and destroyed the military power of the Sassanids' onetime proxies.[2] Thus, when the armies of Islam invaded Syria, they quickly defeated the remaining Lakhmids at Hira in 634. Wheeling right, the Arabs invaded Iraq; turning left, they besieged Damascus.

Heraclius, aged and weakened by decades of battle, nonetheless was alert enough to the danger that he dispatched an army to confront the invaders, which was quickly defeated somewhere between Jerusalem and Gaza. His attention now fully engaged, the emperor assembled a much more formidable force at Antioch, and sent it south.

The size of the forces that would contest the valley of Yarmuk can-

---

* Islam is generally translated as "submission to the will of God"; those who have surrendered are therefore Mu'slam, or Muslim.

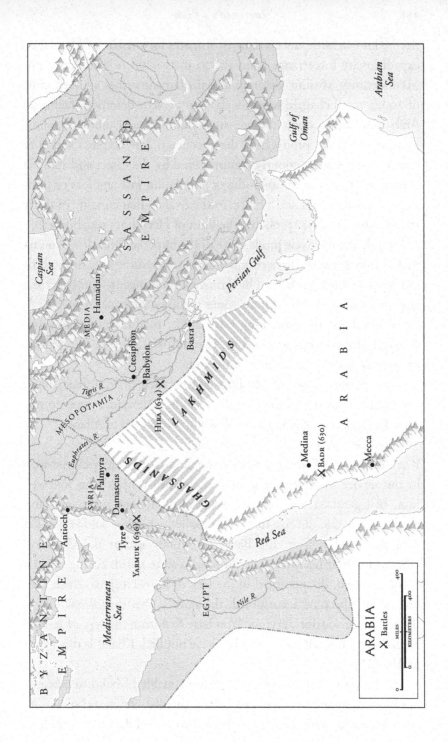

Caspian Sea

Arabian Sea

BYZANTINE EMPIRE

SASSANID EMPIRE

MEDIA
Hamadan

Ctesiphon
Babylon

Tigris R.

MESOPOTAMIA

Euphrates R.

Basra

Gulf of Oman

Persian Gulf

LAKHMIDS

HIRA (634) X

Antioch

SYRIA
Palmyra
Damascus

GHASSANIDS

ARABIA

Medina
X BADR (630)

Mecca

Tyre
YARMUK (636) X

Mediterranean Sea

Red Sea

EGYPT

Nile R.

ARABIA
X Battles

400
MILES
400
KILOMETERS
0

not be known with any certainty. Estimates for the size of Heraclius's expeditionary force range from 15,000 to 150,000 or even more; the ninth-century Muslim historian al-Baladhuri describes a Greek army of 200,000, all chained together so as to prevent desertion, facing an Arab army of 24,000,[3] though such numbers are almost certainly more in service of propaganda than accuracy. The empire's ability to recruit had been so seriously compromised by the direct and indirect consequences of the demon—direct population loss and loss of territory, primarily Illyria and Thrace, to the Avars—that an army of thirty to forty thousand was probably the limit of Heraclius's capabilities.[4]

Despite the relatively modest size of the imperial force, it was nonetheless split between three commanders, representing the multiethnic character of their soldiery. The Armenian contingent—probably the largest—was led by Vahan (in some spellings Mahan), while Jabala ibn ei'Harn led the remaining Ghassanids, in the army's vanguard. The rest of the troops were led by Theodorus Trithurios, but no one, apparently, was in true overall command.

The opposing force of Muslims had no such disadvantage. Their commander was one of the few generals since Alexander never to suffer a defeat in battle: Khalid ibn al-Walid, the Sword of Allah.

––––––––

Khalid arrived at Yarmuk already covered in glory from victories won in the service of Muhammad and of the two caliphs who followed him. In legend, at least, Khalid received his lifelong nickname (and battle cry) in 528, directly from the Prophet, when the soldier had led a fighting retreat from the Battle of Mutah against the Christian Ghassanids after his three superior officers were killed. Facing down a Medina crowd, disappointed in the lack of the victory to which they had quickly become accustomed, the story is that they threw dirt at the returning warriors, crying, "You have fled from the way of Allah." Muhammad himself replied, "They have not fled. Khalid is the Sword of Allah."[5]

His enormous reputation and charisma enabled Khalid to take immediate command of what had been a divided command of the Islamic forces in Syria. His first order was to consolidate them south of

Damascus on the plain of the Yarmuk valley, which today is the border between Jordan and Syria. By the time the Romans arrived, in July of 636, he had selected the ground where he wanted to do battle and placed his troops accordingly, facing west, with their left flank anchored on the ravine overlooking the Yarmuk River. This left the arriving Romans with the option of refusing battle or taking a position facing east, in front of an equally imposing ravine. They had put themselves into a trap: unable to retreat, with ravines to their right and rear.

The Romans may have been trapped, but they were still dangerous. For weeks, both armies avoided serious combat, almost as if the generals knew the significance of their forthcoming confrontation, and felt the weight of history pressing upon them. Moreover, the Roman generals were under orders to pursue negotiations, and offered enormous bribes to Khalid in return for a promised return to Arabia. Between the skirmishes and parleys, it was August 17 before the battle was joined.

For three days, from August 17 to 19, each army proved able to repulse the other's attacks; the ravines of the plateau strongly favored the defense, particularly so long as they had archers. In some accounts, a great wind from the east created a storm that drove sand into the eyes of the Roman line, permitting a decisive attack by Khalid's troops, who routed the Roman cavalry, and rode down the unprotected infantry. Though the legend is dubious—no Arab chronicler records it, and such a poetically appropriate gift to the desert warriors practically demands skepticism—*something* happened to force the issue, whether a sandstorm or a more prosaic turning of the Roman flank. Whatever the cause, on the morning of August 20, the empire lost Syria, one of its oldest and richest provinces, forever. A ninth-century Arab chronicler describes Heraclius leaving Antioch for Constantinople, bidding farewell to his onetime province with the words "Peace unto thee, O Syria, and what an excellent country this is for the enemy."[6]

The conquests that began with the empire's Syrian provinces accumulated thereafter at a frightening pace. The armies of Islam defeated the

Persians at the Battle of Qadesiya in 637, captured Jerusalem in 638, Ctesiphon and Mosul in 641, and Alexandria in 642. Kabul fell in 664, Carthage in 698. Arab armies crossed both the Hindu Kush into Sind in 708 and the Straits of Gibraltar* into Spain in 711. By 721, they had conquered half of Cyprus, most of Sicily, and had crossed the Pyrenees, destroying the last of the Visigoth kingdom.

Constantinople was able to hold out, its geography protecting it, as were the European nations still forming to its north and west, against the Muslim tide. The city of Constantine survived two sieges, in 673–678 and 717–718, and stood unconquered until the cannons of the Ottoman Turks finally breached the Theodosian Walls in 1453. It was not so well guarded against Christians. In 1204, the city was sacked by crusaders, who were by then less representatives of a European super-state called Christendom than of its component nations: Germans, French, English, Venetians, and Hungarians.

A hundred years after Yarmuk, on two battlefields 3,500 miles apart, the Arab armies reached the limits of what would be called the *Dar es Islam* . . . the house of Islam. Though one battlefield witnessed an Arab defeat and the other a victory, both served to define the shape of the next five hundred years.

The first battle is familiar to most historically literate readers; in 732, the outnumbered but disciplined Frankish infantry of Charles Martel† defeated the Arab army that had overwhelmed Africa, Spain, and Bordeaux. In the other, less than twenty years later, on the banks of Central Asia's Talas River, the Arab troops of Ziyad ibn Salih routed an army led by the T'ang general Kao Hsien-Chih. These two battles, Tours and Talas, have an obvious relevance to the question with which this book began: why China remained a coherent nation at precisely the same time that Rome's empire was atomizing into the nations of Europe.

The traditional answer—that Europe (or, more precisely, Christendom) defined itself in opposition to the Islamic nation against which

---

* Gibraltar takes its name from the army's commander, a North African named Tariq: *Jebel Tariq*, or Tariq's Mountain.

† Martel, or Hammer, was the name he earned from his exploits at Tours.

it was locked in mortal combat—is tempting, but unsatisfying. The clash of mutually inimical monotheistic civilizations may have some resonance in the twenty-first century, but projecting it backward into the eighth century is just plain wrong. Neither Tours nor Talas was prompted by an ineradicable conflict between Christian, Muslim, and Buddhist/Confucian. The invasion of southern France that preceded Tours was prompted not by pressure to bring an even larger share of the world under the *shari'a* law of Islam, but by the more mundane desire of the Yemenite governor of al-Andalus to punish the rebellious Berber emir of Catalonia, and to do so by attacking the Berber's ally, the Duke of Aquitaine. Similarly, the great battle between the armies of the Umayyad caliph and the T'ang emperor somewhere near modern-day Kazakhstan was more geopolitics than *jihad*, an opportunistic grab made possible by a revolt by ambitious Tibetans that distracted the Chinese.

And, just as the battles themselves were products of specific circumstances, so, too, were their subsequent historical reputations retrospective creations—as, indeed, is so much else that is usually remembered as evidence of those clashing civilizations. While the *Song of Roland*, the epic poem celebrating Charlemagne's famous knight, memorializes the famous last stand against the Moors at Roncesvaux, the actual rearguard action of the Franks was fought not against Arabs, but Basques. Only centuries later, with the Crusades looming over Europe, were the locals transmuted into examples of the Saracen enemy.

And, even if one grants the importance of Islam in the birth of Christendom, it does nothing to explain why Christendom became France, Spain, Britain, and its other component nations while China, though so soundly defeated at Talas that it was permanently evicted from Central Asia (which remains Islamic to this day), remained China. In the end, then, the most prominent event occurring at the point where the parallels between Chinese and European history diverge is not Islam, but the plague—Justinian's flea.

Which is the second question with which this book began: the importance of the pandemic itself. One way to evaluate the significance of the sixth-century plague—to weigh Justinian's flea—is to examine any subsequent century, asking whether its most prominent events would have occurred at all, or in the same form, in the absence of the pandemic. Consider the following:

Muhammad, for example, would still have received his revelation even had *Y. pestis* never emerged out of Africa, but would Persia and Rome have succumbed so easily had they not lost tens of millions to the bacterium? Had Justinian's successors been able to retain their kingdom on Spain's Mediterranean coast—they had done so, after all, until 625—or Belisarius's conquests in Libya, and so remain athwart the Muslim advance across North Africa; one can only wonder whether the caliphate would have been stymied before ever invading France.

And, even if the armies of Islam would have defeated the two great empires of antiquity anyway, would the Franks have become the pre-eminent power of the ninth century without Charlemagne's acquisition of the title of Holy Roman Emperor? The imperial crown was, after all, granted by a pope who wanted Frankish protection from the Lombards, since he could no longer count on a plague-weakened empire to defend Italy.

Were the Franks to have achieved political leadership in any case, would they have become commercially dominant? Michael McCormick argues, in fact, that during the eighth and ninth centuries, trade with the east no longer overwhelmingly consisted of gold for silk, but of slaves for spice: a direct result of a plague-caused labor shortage in the caliphate.[7]

Even more telling, perhaps, that trade was dominated by the Venetians, residents of a city that did not even exist until 628, as the lagoon became refuge from both plague and Lombards.*

In the end, an appreciation of the significance of Justinian's flea will depend on one's perspective on the importance of Europe, since the demon not only midwifed Europe's birth but fed the population explosion that made the continent the center of historical gravity for a

---

* It is also Venice to whom we owe the word *quarantine*, from the city's custom of restricting ships from port for forty days, a 1377 response to another visit from the demon.

millennium: no pandemic, no labor shortage; no labor shortage, no agricultural revolution; and, therefore, no victory to Europe in the race for population dominance. (Not to forget the other bequest from Justinian to Europe's future dominance: the legal code that legitimized the autocratic European nation-state, and the great landed estates whose taxes paid for it.)

No Holy Roman Empire. No Crusades. No Hundred Years' War. No Inquisition. No European colonies. No Charlemagne, Napoleon, or Hitler. It is not, of course, necessary to imagine a Roman Empire surviving into the twentieth century, as in some alternative history speculative fiction. It is just as intriguing to consider the impact on subsequent events if the empire had been able to hold onto Justinian's conquests for another century or two: had the Mediterranean, like China, been spared a full encounter with *Y. pestis* until the fourteenth century.

The best epidemiological estimates of the recurrent waves of the bubonic plague in late antiquity place the number at eighteen, and while thirteen of them arrived after the Battle of Yarmuk, almost all stayed east of Constantinople.[8] The first waves of plague—in 542, 558, 590, 597, and 618—are the ones that sabotaged Justinian's reconquest, and permitted the European states to survive their infancies.

Or, perhaps not. The forces acting on the formation of modern Europe are impossible to number, much less quantify. There is no unique solution to the historical problem, which is another way of admitting that Europe might look precisely the same even had *Y. pseudotuberculosis* never made those few hundred changes in its genome; the three thousand-body problem remains unsolved.

It can, however, be illuminated. One path to such illumination is straightforward, the belief that historical change is linear, a measured accumulation of small changes adding up to huge transformations over time. Plate tectonics, classical evolutionary biology, and long-wave historiography are all linear systems. The other path is a series of switchbacks, a historical record built on catastrophic discontinuities. Examples are extinctions, asteroid impacts, tidal waves—and plagues.

A tidal wave leaves a mark by the force of its collision with the shore, but also by carving new channels in the land over which it washes and then recedes. The most long-lasting effect of the plague was not its initial impact, but the way in which its aftershocks remade the topography of Europe and the Mediterranean. As the demon washed across the lands once ruled by Rome, it left behind tidal pools: the distinctive regions in which protonations like the Franks, Lombards, Saxons, Slavs, and Goths could coalesce and combine into polities called France, Spain, and England . . . and, a few centuries later, into Italy, Germany, and the Netherlands. The consequence—the birth of Europe—is what prompted Josiah Russell, forty years ago, to write: "In the whole Mediterranean-European complex, neither Charlemagne, nor Harun [el-Rashid] nor the great Isaurian and Macedonian dynasties could break the pattern set up by the flea, the rat, and the bacillus."[9]

I N THE SPRING *of 1347, a Black Sea port city and Genoese colony, Caffa, was under siege. The invading army consisted of Crimean Mongols and soldiers from Venice . . . like its bitter rival Genoa, one of the Italian city-states that formed in the peninsula as imperial control receded. In October of that year, a fleet of trading ships fleeing the besieged city arrived in the Sicilian port of Messina. All the crew members were either infected or dead, most decorated with the grapefruit-sized swellings called* buboes.

Like the comet that heralded its first arrival eight hundred years before, the demon, this time called the Great Mortality or the Black Death, had returned.

ACKNOWLEDGMENTS

After several decades of editing and publishing books by others, one acquires some unfortunate habits, such as reading a book's acknowledgments page first, sometimes to acquire professional intelligence, sometimes to decode the ways in which authors express gratitude. Until writing *Justinian's Flea*, however, I never realized how pleasurable it is to finally thank the people upon whose help every writer depends.

A book like this is not even imaginable without access to the resources of a first-rate research library, and Princeton University is certainly rich in them. Among the university's many assets are the people who staff those libraries, one of whom was among my first and most important guides, Elizabeth Bennett of Princeton's Firestone Library. Also at Princeton, Kathy Daumer and Andrew McCandlish of the biology department, and Kutlu Akalin and Craig Caldwell of the history department, were utterly invaluable in both directing me to other resources and in correcting an embarrassing number of misapprehensions about everything from cell biology to Monophysitism. I hasten to add that they bear no responsibility for the errors that have crept in since last they saw the work.

A number of scholars of the period were likewise both helpful and encouraging, like Dionysios Stathakopoulos, Michael Morony, John Maddicott, Lester B. Little, and David Whitehouse. As above, whenever I have been accurate and/or perceptive in this book, I have certainly been channeling the work of others; inaccuracies and infelicities are owned by me alone.

As a onetime editor myself, I have reminded authors beyond counting that everyone needs editing. I have been both happy and fortunate to be reminded of it myself, this time by Will Sulkin at Jonathan

Cape, and especially Rick Kot at Viking, men of skill, taste, and great kindness. Their respective staffs—Rosalind Porter at Cape; Alessandra Lusardi and Laura Tisdel at Viking (as well as Jennifer Tait, who supervised the book's production)—have been unfailingly helpful and intimidatingly efficient. Carla Bolte, who designed the book, and David Lindroth, who created the maps, have reminded me how even the clumsiest prose can be made to seem elegant, if one can only find talented people who know that books appeal to the eye as well as the mind. I am in debt to them both.

In my publishing houses, as in much else connected with the writing of *Justinian's Flea*, I have been extraordinarily fortunate, but the good fortune that has been one of the most distinctive characteristics of this project began with the world's best agents, Eric Simonoff and Cecile Barendsma at Janklow & Nesbit, the first readers of this book at every stage in its creation, and its best advocates. I am grateful beyond words to both of them.

Thanks, too, to family and friends who were kind enough to read the manuscript, evaluate jacket designs, and even weigh in on titles, including David Jacobus; Tom Keenan; Holly Goldberg Sloan; Ann Margaret Daniel; my brother Gary, my children Quillan, Emma, and Alex; and—of course—to my wife, Jeanine, the woman who asked me the question "what would you do if you were unafraid to fail?" for which the answer became *Justinian's Flea*.

NOTES

Introduction: The Three Thousand-Body Problem

1. Anthony Pagden, ed., *The Idea of Europe: From Antiquity to the European Union* (Cambridge: Cambridge University Press, 2002).

Prologue: Pelusium

1. A. H. M. Jones, *The Later Roman Empire, 284–602* (Baltimore: Johns Hopkins University Press, 1964).

1. "Four Princes of the World"

1. Virgil, *The Aeneid*, trans. John Dryden, edited, with introduction and notes, by Robert Fitzgerald (New York: Macmillan, 1965).
2. Procopius, *History of the Wars*, trans. H. B. Dewing (Cambridge, MA: Harvard University Press, 1914).
3. Cyril Mango, *Byzantium, The Empire of New Rome* (New York: Scribner's, 1980).
4. John Haldon, "Humour and the Everyday in Byzantium" in *Humour, History, and Politics in the Late Antiquity and the Early Middle Ages*, Guy Halsall, ed. (Cambridge: Cambridge University Press, 2002).
5. Cyril Mango, "The Triumphal Way of Constantinople and the Golden Gate," *Dumbarton Oaks Papers*, no. 54, 2000.
6. A. H. M. Jones, *The Later Roman Empire, 284–602* (Baltimore: Johns Hopkins University Press, 1964).
7. Ibid.
8. Ibid.
9. James Allen Evans, "Introduction to the Early Byzantine: Constantinople and the Basilica of Hagia Sophia," *Athena Review*, vol. 3, no. 1, 1996.
10. Robert Browning, *Justinian and Theodora* (New York: Praeger, 1971).
11. Jones, *The Later Roman Empire.*
12. Ibid.
13. Ibid.
14. Felipe Fernandez-Armesto, *The Times Illustrated History of Europe* (London: Times Books, 1995).
15. Browning, *Justinian and Theodora.*
16. J. J. Wilkes, *Diocletian's Palace, Split: Residence of a Retired Roman Emperor,* Department

of Ancient History and Classical Archaeology Occasional Publications, no. 2, University of Sheffield.

17. Edward Gibbon, *The Decline and Fall of the Roman Empire* (New York: E. P. Dutton, 1952).

18. M. J. Nicasie, *Twilight of Empire: The Roman Army from the Reign of Diocletian Until the Battle of Adrianople* (Amsterdam: J. C. Giegen, 1998). The *Notitia Dignitatum* of the late fourth century "yields the staggering number of more than 200 legions."

19. Jones, *The Later Roman Empire*.

20. Agathias, *The Histories*, vol. 13, trans. Joseph D. Frendo (New York: De Gruyter, 1975).

21. Nicasie, *Twilight of Empire*.

22. Gibbon, *Decline and Fall*, ch. 13.

23. Ibid., ch. 18.

24. Ibid., ch. 14.

25. Louis Duchesne, *Early History of the Christian Church* (New York: Longmans Green, 1912–24).

26. Eusebius, *Ecclesiastical History*, trans. Kirsopp Lake (Cambridge, MA: Harvard University Press, 1984). Quoted in Evans, *Procopius*.

27. Michael Whitby and Mary Whitby, trans., *Chronicon Paschale 284–628 AD* (Liverpool: Liverpool University Press, 1989).

28. Ibid.

29. Pierre Maraval, "The Earliest Phase of Christian Pilgrimage in the Near East," *Dumbarton Oaks Papers*, no. 56, 2003. A number of third-century visitors precede Helena, including Origen of Alexandria, who "visited some sites in search of traces of Jesus." Constantine's mother, however, was explicitly on a mission from Bishop Makarios of Jerusalem to seek out the sites of the three key moments in the historical life of Jesus: his birth at Bethlehem, his death at Golgotha, and his resurrection on the Mount of Olives.

30. Procopius, *The Secret History*, trans. G. A. Williamson (New York: Penguin Books, 1966); Tony Honore, *Tribonian* (London: Duckworth Publishing, 1978).

31. Michael Maas, "Roman Questions, Byzantine Answers" in *The Cambridge Companion to the Age of Justinian* (New York: Cambridge University Press, 2005), quoting Kroll's translation of Justinian's *Pragmatic Sanction*.

2. "We Do Not Love Anything Uncivilized"

1. Herwig Wolfram, *History of the Goths*, trans. Thomas J. Dunlap (Berkeley: University of California Press, 1988).

2. Charles Christopher Mierow, trans., *The Gothic History of Jordanes* (Princeton: Princeton University Press, 1915).

3. M. J. Nicasie, *Twilight of Empire: The Roman Army from the Reign of Diocletian Until the Battle of Adrianople* (Amsterdam: J. C. Giegen, 1998).

4. Wolfram, *History of the Goths*.

5. Ibid.

6. Ammianus Marcellinus, *The Later Roman Empire, (354–378)*, trans. Walter Hamilton (New York: Penguin Books, 1986).

7. Ibid.

8. John Keegan, *A History of Warfare* (New York: Knopf, 1993).

9. Marcellinus, *The Later Roman Empire*.

10. Nicasie, *Twilight of Empire*. Some have estimated as many as 200,000.

11. Marcellinus, *The Later Roman Empire*.

12. Nicasie, *Twilight of Empire*. No one knows the size of either army with great precision. Estimates range from the 20,000 given above to as much as 45,000, or even 60,000. Similarly, the Goths might have been as few as 10,000.

13. Wolfram, *History of the Goths*.

14. Marcellinus, *The Later Roman Empire*.

15. Nicasie, *Twilight of Empire*.

16. Noel Lenski, *The Cambridge Companion to the Age of Constantine* (New York: Cambridge University Press, 2005).

17. Ibid.

18. Marcellinus, *The Later Roman Empire*.

19. Wolfram, *History of the Goths*.

20. Procopius, *History of the Wars*, III, 1, trans. H. B. Dewing (Cambridge, MA: Harvard University Press, 1914).

21. Wolfram, *History of the Goths*.

22. Edward Gibbon, *The Decline and Fall of the Roman Empire*, ch. 30 (New York: E. P. Dutton, 1952).

23. A. H. M. Jones, *The Later Roman Empire, 284–602* (Baltimore: Johns Hopkins University Press, 1964).

24. Mierow, *Jordanes*.

25. Ibid.

26. Wolfram, *History of the Goths*.

27. Jones, *The Later Roman Empire*. Jones cites the census taken by Gaiseric prior to the crossing into Africa as reported by Victor Vitensis.

28. Marcellinus, *The Later Roman Empire*.

29. R. C. Blockley, *The Fragmentary Classicizing Historians of the Later Roman Empire: Eunapius, Olympiodorus, Priscus, and Malchus*, vol. II, trans. J. B. Bury (Liverpool: Francis Cairns, 1983).

30. Richard Gordon, "Stopping Attila: The Battle of Chalons," *Military History*, December 2003.

31. Blockley, *Historians of the Later Roman Empire*.

32. Jones, *The Later Roman Empire*.

33. Blockley, *Historians of the Later Roman Empire*.

34. Peter Heather, "The Huns and the End of the Roman Empire in Western Europe," *English Historical Review*, February 1995.

35. Jones, *The Later Roman Empire*.

36. Kelley Ross, "Rome and Romania," www.friesian.com/romania.htm.

37. Jones, *The Later Roman Empire*.

38. Wolfram, *History of the Goths*.

39. Thomas Hodgkin, trans. *Letters of Cassiodorus* (London: H. Frowde, 1886).

40. Pope Leo I, "Letter 28.4," in *Creeds, Councils, and Controversies* (London: SPCK Publishing, 1966).

41. John Henry Cardinal Newman, *Tracts Theological and Ecclesiastic*, iv, 1874.

42. O. Guenther, ed., *Collection Avellana* (*Corpus Scriptorum Ecclesiasticorum, Latinorum*, vol. 35; Vienna, 1895), quoted in Bury.

43. Jones, *The Later Roman Empire*.

3. "Our Most Pious Consort"

1. Peter Brown, *The World of Late Antiquity* (London: Thames & Hudson, 1971). This is sometimes translated as "An impoverished Roman imitates a Goth, and a rich Goth imitates a Roman."

2. J. B. Bury, *History of the Later Roman Empire*, vol. 2 (New York: Dover, 1958).

3. Anicius Boethius, *The Consolation of Philosophy*, trans. V. E. Watts (New York: Penguin Books, 1980).

4. Felipe Fernandez-Armesto, *Times Illustrated History of Europe* (London: Times Books, 1995).

5. Bury, *History of the Later Roman Empire*.

6. Procopius, *The Secret History*, trans. G. A. Williamson (New York: Penguin Books, 1966).

7. Evagrius Scholasticus, *Ecclesiastical History*, trans. E. Walford (London: H. G. Bohn, 1846).

8. Paul Krueger, *The Digest of Justinian*, trans. Alan Watson (Philadelphia: University of Pennsylvania Press, 1985).

9. Robert Browning, *Justinian and Theodora* (New York: Praeger, 1971).

10. Ibid.

11. A. H. M. Jones, *The Later Roman Empire, 284–602* (Baltimore: Johns Hopkins University Press, 1964).

12. Alan Cameron, *Circus Factions: Blues and Greens at Rome and Byzantium* (New York: Oxford University Press, 1993).

13. Jones, *The Later Roman Empire*.

14. Theophanes, *Chronicle*, quoted in J. B. Bury, *History of the Later Roman Empire*.

15. Jones, *The Later Roman Empire*.

16. Scholasticus, *Ecclesiastical History*.

17. Browning, *Justinian and Theodora*.

18. Procopius, *The Secret History*, 9.20.

19. Tony Honore, *Tribonian* (London: Duckworth Publishing, 1978).

20. Browning, *Justinian and Theodora*.

21. Procopius, *History of the Wars*, trans. H. B. Dewing (Cambridge, MA: Harvard University Press, 1914), III, 10.

22. Browning, *Justinian and Theodora*.

23. Ibid.

24. John L. Teall, "The Barbarians in Justinian's Armies," *Speculum*, vol. 40, 1965.

25. Ibid.

26. Bury, *History of the Later Roman Empire*.

27. Erik Hildinger, "Belisarius's Bid for Rome" in *Military History*, October 1999.

28. Procopius, *Wars*, VI, 14.

29. Lawrence Fauber, *Narses: Hammer of the Goths* (New York: St. Martin's, 1990).

30. Procopius, *Wars*, I, 3.

31. Ibid, VI, 14.

4. "Solomon, I Have Outdone Thee"

1. The *Chronicle* of John Malalas says of Justinian, "He established a secure, orderly condition in every city of the Roman state and dispatched sacred rescripts to every

city so that rioters or murderers, no matter to what faction they belonged, were to be punished." Geoffrey Greatrex, "The Nika Riot: A Reappraisal," *Journal of Hellenic Studies*, 1997.

2. J. B. Bury, *History of the Later Roman Empire*, vol. 2 (New York: Dover, 1958).

3. A. H. M. Jones, *The Later Roman Empire, 284–602* (Baltimore: Johns Hopkins University Press, 1964).

4. Bury, *History of the Later Roman Empire.*

5. Ibid.

6. Procopius, *History of the Wars*, I, 24, 37, trans. H. B. Dewing (Cambridge, MA: Harvard University Press, 1914).

7. Henry Adams, *Mont Saint Michel and Chartres* (New York: Library of America, 1983).

8. Georgio Cedrenus, *Compendium Historiarum*, in R. J. Mainstone, *Hagia Sophia: Architecture, Structure, and Liturgy of Justinian's Great Church* (London: Thames & Hudson, 1988).

9. Mainstone, *Hagia Sophia.*

10. Eusebius, *Vita Constantina*, vol. 20, trans. E. C. Richardson, 1890, quoted in Mainstone, *Hagia Sophia.*

11. A. K. Orlandos, *Byzantine Architecture* (Athens: Archaeological Society at Athens, 1998).

12. Mainstone, *Hagia Sophia.*

13. Adams, *Mont St. Michel and Chartres.*

14. Mainstone, *Hagia Sophia.*

15. John Warren, "Greek Mathematics and the Architects to Justinian," *Art and Archaeology Research Papers* (London: Coach Publishing, 1976).

16. George Huxley, *Anthemius of Tralles and Later Greek Geometry.*

17. Ilhan Aksit, *Hagia Sophia and Kariye Museum* (Istanbul, nd).

18. Mainstone, *Hagia Sophia.* Mainstone suggests that a few lines in Agathias's *Historia*, written long after the building of the church, describe some additional elements for the original brief, most particularly the requirement that the church be built without wood, in order to make it as fireproof as possible.

19. Procopius, *Buildings*, 1.1 (Cambridge, MA: Loeb Classical Library, 1940).

20. Pero Tafur, *Travels and Adventures* (c. 1455), quoted in John Balfour, Baron Kinross, *Hagia Sophia* (New York: W. W. Norton & Co., 1972).

21. Mainstone, *Hagia Sophia.*

22. Procopius, *Buildings*, 1.1.

23. Bury, *History of the Later Roman Empire*, vol. 2.

24. Orlandos, *Byzantine Architecture.*

25. Paul the Silentiary, quoted in Mainstone, *Hagia Sophia.*

26. Mainstone, *Hagia Sophia.*

27. Ibid.

28. Procopius, *Buildings*, 1.1.

29. Ibid.

30. Mainstone, *Hagia Sophia.*

31. Warren, "Greek Mathematics and the Architects to Justinian."

32. Ibid.

33. Guillermo Rivoira, *Roman Architecture and Its Principles of Construction under the Empire* (Oxford: Clarendon Press, 1925).

34. Agathias, *The Histories,* vol. 8, trans. Joseph D. Frendo (New York: DeGruyter, 1975).
35. Paul the Silentiary, quoted in Lord Kinross, *Hagia Sophia.*
36. Mainstone, *Hagia Sophia.*
37. Evagrius Scholasticus, *Ecclesiastical History,* bk. 4, trans. E. Walford (London: H. G. Bohn, 1846).
38. Agathias, *Histories,* vol. 8.

5. *"Live Honorably, Harm Nobody, and Give Everyone His Due"*

1. From second-century Syrian Christian apologist Tatian in his *Oratio ad Graecos,* in Peter Brown, *The World in Late Antiquity* (London: Thames & Hudson, 1971).
2. Tony Honore, *Tribonian* (London: Duckworth Publishing, 1978).
3. Ibid.
4. Caroline Humfress, "Law and Legal Practice in the Age of Justinian," Michael Mass, ed., *The Cambridge Companion to the Age of Justinian* (Cambridge: Cambridge University Press, 2005).
5. Procopius, *History of the Wars,* 1.24. 16, trans. H. B. Dewing (Cambridge, MA: Harvard University Press, 1914).
6. Honore, *Tribonian.*
7. Peter Birks and Grant McLeod, trans., *Justinian's Institutes* (Ithaca, NY: Cornell University Press, 1987).
8. Quoted in Honore, *Tribonian.*
9. Justinian, *Constitution of 528,* quoted in Robert Browning, *Justinian and Theodora* (New York: Praeger, 1971).
10. Honore, *Tribonian.*
11. Ibid.
12. Ibid.
13. A. P. d'Entreves, *Natural Law: An Introduction to Legal Philosophies* (London: Hutchinson's University Library, 1951), quoted in Justinian, *The Digest of Roman Law* (New York: Penguin, 1989).
14. Will Durant, *The Age of Faith* (New York: Simon & Schuster, 1950).
15. Justinian, *Digest of Roman Law,* trans. C. F. Kolbert (New York: Penguin Books, 1979).
16. E. Stein, *Histoire de Bas-Empire* (Paris: 1959).
17. Henry John Stephen, "A Treatise on the Principles of Pleading in Civil Actions: Comprising a Summary View of the Whole Proceedings in a Suit at Law" (Washington, D.C.: Walter C. Morrison Publishers, 1898).
18. Charles Casassa, "Magister Vacarius 'Hic en Oxonefordia Legem Docuit': An Analysis of the Dissemination of Roman Law in the Middle Ages." Online, http://history.eserver.org/dissemination-of-law.txt.
19. Stephen, "A Treatise on the Principles of Pleading in Civil Actions."
20. Joseph Story, *Commentaries on the Conflict of Laws, Foreign and Domestic, in Regard to Contracts, Rights, and Remedies, and Especially in Regard to Marriages, Divorces, Wills, Successions, and Judgments* (London, 1834), quoted in Stephen, "A Treatise on the Principles of Pleading in Civil Actions."
21. Honore, *Tribonian.*
22. R. Schoell and G. Kroll, eds. *Institutiae Novellae* from the *Corpus Iuris Civilis* vol. iii, 6th ed. (Berlin: Apud Weidmannos, 1954).

6. "The Victories Granted Us by Heaven"

1. R. Schoell and G. Kroll, eds., *Institutiae Novellae* from the *Corpus Iuris Civilis* vol. iii, 6th ed. (Berlin: Apud Weidmannos, 1954).
2. Stephen Muhlberger, "Overview of Late Antiquity—The Fifth Century," Section 2: "Weak Emperors and Warlords" ORB Online Encyclopedia, http://www.nipissingu.ca/department/history/MUHLBERGER/ORB/OVC3S2.HTM).
3. A. H. M. Jones, *The Later Roman Empire, 284–602* (Baltimore: Johns Hopkins University Press, 1964). Jones cites the census taken by Gaiseric prior to the crossing into Africa as reported by Victor Vitensis.
4. Procopius, *History of the Wars,* III, 3, trans. H. B. Dewing (Cambridge, MA: Harvard University Press, 1914).
5. Ibid.
6. Ibid., III. 4.
7. Jacques Le Goff, *Medieval Civilization, 400–1500* (Oxford: Blackwell Publishers, 1988).
8. Procopius, *Wars,* III, 5.
9. Ibid., III, 9.
10. Ibid.
11. Jones, *The Later Roman Empire.*
12. Edward Gibbon, *The Decline and Fall of the Roman Empire* (New York: G. P. Dutton, 1952).
13. Procopius, *Wars,* III, 21.
14. Robert Browning, *Justinian and Theodora* (New York: Praeger, 1971); Procopius, *Wars,* III, 25.
15. Peter Birks and Grant McLeod, trans., *Justinian's Institutes* (Ithaca, NY: Cornell University Press, 1987).
16. Procopius, *Wars,* IV, 10.20; Cyril Mango, "The Triumphal Way of Constantinople and the Golden Gate," *Dumbarton Oaks Papers,* no. 54, 2000.
17. Jones, *The Later Roman Empire,* vol. 2.
18. Bernard Lewis, *The Middle East* (New York: Scribner, 1995).
19. Gershon Cohen, "The Talmudic Age" in *Great Ages and Ideas of the Jewish People,* Leo Schwarz, ed. (New York: Modern Library, 1956).
20. Stefan Zweig, *The Buried Candelabrum* (New York: Viking Press, 1937).
21. Nicholas de Langue, "Jews in the Age of Justinian," Michael Maas, ed., *The Cambridge Companion to the Age of Justinian* (Cambridge: Cambridge University Press, 2005).
22. Procopius, *Wars,* IV, 5.
23. Browning, *Justinian and Theodora.*
24. Gibbon, *Decline and Fall,* ch. 41.
25. Schoell and Kroll, *Institutae Novellae.*
26. Procopius, *Wars,* VIII, 22.7.
27. Erik Hildinger, "Belisarius's Bid for Rome," *Military History,* October 1999.
28. Procopius, *Wars,* V, 18.
29. Ibid., V, 19.
30. Hildinger, "Belisarius's Bid for Rome."
31. Procopius, *Wars,* V, 24.

32. Ibid., *Wars,* VI, 6.

33. Ibid., *Wars,* VI, 10.

34. Ibid.

35. Hildinger, "Belisarius's Bid for Rome."

36. Procopius, *Wars,* VI, 10.

37. Lawrence Fauber, *Narses: Hammer of the Goths* (New York: St. Martin's, 1990).

38. Agathias, *The Histories,* trans. Joseph D. Frendo (New York: De Gruyter, 1975).

39. Gibbon, *Decline and Fall.*

40. Fauber, *Narses.*

41. Procopius, *Wars,* VI, 16.

42. Ibid., VI, 20.

43. Ibid., VI, 28.

44. Ibid., V, 1.

45. Annabel Jane Wharton, "Ritual and Reconstructed Meaning: The Neonian Baptistery in Ravenna," *The Art Bulletin,* vol. 69, no. 3, 1987.

46. UNESCO, *Justification for Inclusion in the World Heritage List.*

47. Procopius, *Buildings* 1.2 (Cambridge, MA: Loeb Classical Library, 1949).

48. Cyril Mango, "The Triumphal Way of Constantinople and the Golden Gate," *Dumbarton Oaks Papers* no. 54, 2000.

49. Procopius, *Wars,* VI, 29.

50. J. B. Bury, *History of the Later Roman Empire* (New York: Dover, 1958).

7. *"Daughter of Chance and Number"*

1. Lynn Margulis and Dorian Sagan, *What Is Life?* (New York: Simon & Schuster, 1995).

2. Charles Davenant, quoted in Simon Schama, *The Embarrassment of Riches* (New York: Knopf, 1987).

3. S. J. Gould, *Full House* (New York: Harmony Books, 1996).

4. Kenneth Todar, *Online Encyclopedia of Microbiology,* http://www.textbookofbacteriology.net.

5. Clifford Dobell, *Antony van Leeuwenhoek and His "Little Animals"* (New York: Dover, 1978).

6. B. Joseph Hinnebusch et al., "Role of Yersinia Murine Toxin in the Survival of *Yersinia pestis* in the Midgut of the Flea Vector," *Science,* vol. 296, 2002.

7. C. O. Jarrett et al. "Transmission of *Yersinia pestis* from an Infectious Biofilm in the Flea Vector," *Journal of Infectious Diseases,* July 2004.

8. Susan Scott and Christopher Duncan, *Biology of Plagues: Evidence from Historical Populations* (Cambridge: Cambridge University Press, 2001).

9. M. Drancourt, V. Roux, L. V. Dang, L. Tran-Hung, D. Castex, V. Chenal-Francisque, et al. "Genotyping, Orientalis-like *Yersinia pestis,* and plague pandemics." *Emerg. Infect. Dis.,* vol. 10, no. 9, 2004.

10. V. Chenal-Francisque et al., "Insights into the Genome Evolution of *Y. pestis* through Whole Genome Comparison with *Y. pseudotuberculosis,*" *Proceedings of the National Academy of Sciences,* vol. 101, no. 38, 2004.

11. Stuart A. Kauffman, "The Sciences of Complexity and Origins of Order," *PSA: Proceedings of the Biennial Meeting of the Philosophy of Science Association,* Chicago, 1990. Kauffman's favored analogy is not Legos, but buttons and thread.

12. A. L. Burroughs, "Sylvatic Plague Studies: The Vector Efficiency of Nine Species of Fleas Compared with *Xenopsylla Cheopsis,*" *Journal of Hygiene,* vol. 43, 371–396.

13. L. Kartman et al., "New Knowledge of the Ecology of Sylvatic Plague," *Annals of the New York Academy of Science,* vol. 70: 668–711.

14. Todar, *Online Encyclopedia of Microbiology.*

15. Edward O. Wilson, *The Diversity of Life* (Cambridge, MA: Harvard University Press, 1992).

8. *"From So Simple a Beginning"*

1. D. E. Davis, "The Scarcity of Rats and the Black Death: An Ecological History," *Journal of Interdisciplinary History,* XVI, 1986.

2. Michael McCormick, "Rats, Communications, and Plague: Toward an Ecological History," *Journal of Interdisciplinary History,* vol. 34, no. 1.

3. Ibid.

4. E. LeCompte et al., "Integrative Systematics: Contributions to Phylogeny and Evolution," in Grant Singleton et al., eds., *Rats, Mice, and People: Rodent Biology and Management* (Canberra: Australian Centre for International Agricultural Research, 2003).

5. R. Nowak, *Walker's Mammals of the World,* vol. 2, fifth ed. (Baltimore: Johns Hopkins University Press, 1991).

6. McCormick, "Rats, Communications, and Plague."

7. Ibid.

8. Ibid.

9. Nowak, *Mammals of the World.*

10. Wendy Orent, *Plague* (New York: Free Press, 2004).

11. William H. MacNeill, *Plagues and Peoples* (New York: Anchor Books, 1998).

12. Ibid.

13. Ibid.

14. Jared Diamond, *Guns, Germs, and Steel* (New York: W. W. Norton, 1997).

15. M. J. Keeling and C. A. Gilligan, "Metapopulation Dynamics of Bubonic Plague," *Nature,* vol. 207, October 2000.

16. Robert B. Strassler, *The Landmark Thucydides: A Newly Revised Edition of the Robert Crawley Translation* (New York: Free Press, 1996).

17. Richard M. Krause, "The Origin of Plagues, Old and New," *Science,* vol. 257, 1992.

18. McNeill, *Plagues and Peoples.*

19. Malcolm Gladwell, *The Tipping Point* (New York: Little, Brown, 2000).

20. McCormick, "Rats, Communications, and Plague."

21. M. J. Keeling and C. A. Gilligan, "Bubonic Plague," *Proceedings of the Royal Society of London, Biological Sciences,* vol. 267, 2000.

22. McNeill, *Plagues and Peoples.*

23. P. Sarris, "The Justinianic Plague: Origins and Effects," *Continuity and Change,* 17.2, Cambridge, 2002.

24. Ibid.

25. Eva Panagiotakopulu, "Pharaonic Egypt and the Origins of Plague," *Journal of Biogeography,* vol. 31/2, 2004.

26. Evagrius Scholasticus, *Ecclesiastical History*, book 4, trans. E. Walford (London: H. G. Bohn, 1846).

27. David Keys, *Catastrophe: An Investigation into the Origins of Modern Civilization* (New York: Ballantine Books, 2000).

28. H. Pirenne, *Mohammed and Charlemagne* (London: Allen & Unwin, 1939).

9. "The Fury of the Wrath of God"

1. T. H. Huxley, "A Lecture to the Eton Volunteer Corps," *Macmillan's Magazine*, London, 1883.

2. A. A. Vasiliev, trans., *Description of the Entire World* ("Expositio Totius Mundi"), *Seminarium Kondakovium*, 8, 1935.

3. Ibid.

4. Cyril Mango, *Byzantium: The Empire of New Rome* (New York: Scribner's, 1980).

5. Neville G. Brown, *Challenge of Climatic Change* (New York: Routledge, 1999).

6. Mike Baillie, *A Slice Through Time, Dendochronology and Precision Dating* (London: Batsford Publishing, 1995).

7. Nick Nuttall, "Tale of Arthur Points to Comet Catastrophe," *The Times*, September 9, 2000.

8. Procopius, *History of the Wars*, II, 3, trans. H. B. Dewing (Cambridge, MA: Harvard University Press, 1914).

9. Ibid., IV, 14.

10. Richard B. Stothers, "Volcanic Dry Fogs, Climate Cooling and Plague Pandemics in Europe and the Middle East," *Climatic Change*, vol. 42, no. 4, August 1999.

11. A. Blocker, K. Komoriya, et al., "Type III Secretion Systems and Bacterial Flagella: Insights into Their Function from Structural Similarities," *Proceedings of the National Academy of Sciences*, March 18, 2003.

12. Luke A. J. O'Neill, "Immunity's Early-Warning System," *Scientific American*, January 2005.

13. Guy R. Cornelis, "Molecular and Cell Biology Aspects of Plague," *Proceedings of the National Academy of Sciences*, vol. 97, no. 16, August 2000.

14. Wendy Orent, *Plague* (New York: Free Press, 2004).

15. Kenneth Todar, *Online Encyclopedia of Microbiology*, http://www.textbookofbacteriology.net.

16. P. B. Medawar, *Aristotle to Zoos* (Cambridge, MA: Harvard University Press, 1983).

17. Orent, *Plague*.

18. T. H. Hollingsworth, *Historical Demography: The Sources of History*, cited in Dionoysus Stathakopolous, *Famine and Pestilence in the Late Roman and Early Byzantine Empire: A Systematic Survey of Subsistence Crises and Epidemics* (London: Ashgate Publishing, 2004).

19. Peregrine Horden, "Mediterranean Plague in the Age of Justinian" in Michael Maas, *The Cambridge Companion to the Age of Justinian* (Cambridge: Cambridge University Press, 2005).

20. Procopius, *Wars*, II, xxii.8.

21. A. H. M. Jones, *The Later Roman Empire, 284–602* (Baltimore: Johns Hopkins University Press, 1964).

22. John Duffy, "Byzantine Medicine in the Sixth and Seventh Centuries: Aspects of Teaching and Practice," *Dumbarton Oaks Papers*, no. 38, 1984.

23. Ibid.

24. John Scarborough, "Early Byzantine Pharmacology," *Dumbarton Oaks Papers*, no. 38, 1984.

25. Vivian Nutton, "Galen to Alexander: Medical Practice in Late Antiquity," *Dumbarton Oaks Papers*, 1984.

26. Gary Vikan, "Art, Medicine, and Magic in Early Byzantium," *Dumbarton Oaks Papers*, no. 38, 1984.

27. Isidore of Seville, *On Medicine*, trans. William D. Sharpe, *Transactions of the American Philosophical Society*, vol. 54, 1964.

28. Scarborough, "Early Byzantine Pharmacology."

29. Jerry Stannard, "Aspects of Byzantine Materia Medica," *Dumbarton Oaks Papers*, no. 38, 1984.

30. Lewis Thomas, *The Fragile Species* (New York: Scribner, 1992).

31. Duffy, "Byzantine Medicine in the Sixth and Seventh Centuries."

32. Nutton, "Galen to Alexander."

33. Ibid.

34. Timothy S. Miller, "Byzantine Hospitals," *Dumbarton Oaks Papers*, no. 38, 1984.

35. J. N. Biraben and Jacques Le Goff, "The Plague in the Early Middle Ages" in *Biology of Man in History*, edited and translated by Robert Forster and Orest Ranum, trans. and eds. (Baltimore: Johns Hopkins University Press, 1975).

36. Procopius, *Wars*, II, 23.

37. Edward Gibbon, *Decline and Fall of the Roman Empire*, ch. 40 (New York: G. P. Dutton, 1952).

38. Procopius, *Wars*, II, 21.

39. Ibid., II, 22.

40. A. B. Christie, *Infectious Diseases: Epidemiology and Clinical Practice*, 3rd ed. (Edinburgh: Churchill Livingstone, 1969). Quoted in Susan Scott and Christopher Duncan, *Biology of Plagues: Evidence from Historical Populations* (Cambridge: Cambridge University Press, 2001). Note that Scott and Duncan make much of the fact that Christie himself observes that many of these symptoms are those of other severe illnesses, such as malaria, typhus, hemorrhagic viral fevers, and so on. They write that compared with these other diseases, *"the only distinguishing feature is the bubo."* This is an important point for their thesis regarding the Black Death, but does nothing to suggest that Justinian's Plague was not caused by *Y. pestis,* given the well-documented appearance of the buboes.

41. Evagrius Scholasticus, *Ecclesiastical History*, trans. E. Walford (London: H. G. Bohn, 1846).

42. Cyprian, *De Mortalitate*, trans. Mary Louise Hannon, quoted in McNeill, *Plagues and Peoples*.

43. Andrew Palmer and Sebastian Brock, trans., *Apocalypse of Pseudo-Methodius*, in *The Seventh Century in the West-Syrian Chronicle* (Liverpool: Liverpool University Press, 1993).

44. Susan Ashbrook Harvey, "Physicians and Ascetics in John of Ephesus: An Expedient Alliance," *Dumbarton Oaks Papers*, no. 38, 1984.

45. Witold Witakowski, *The Chronicle of Pseudo-Dionysius of Tel-Mahre*, Part III (Liverpool: Liverpool University Press, 1997).
46. Ibid.
47. Harvey, "Physicians and Ascetics."
48. Witakowski, *Chronicle of Pseudo-Dionysus*.
49. Ibid.
50. Ibid.

10. "A Man of Unruly Mind"

1. J. B. Bury, *History of the Later Roman Empire*, v. 2 (New York: Dover, 1958).
2. David Kennedy and Derrick Riley, *Rome's Desert Frontier* (Austin: University of Texas Press, 1990).
3. Ibid.
4. Elton L. Daniel, *History of Iran* (New York: Greenwood Publishing, 2000).
5. Ammianus Marcellinus, *The Later Roman Empire (354–378)*, Walter Hamilton, trans. (New York: Penguin Books, 1986).
6. Mary Boyce, *Zoroastrians, Their Religious Beliefs and Practices* (London: Routledge and Kegan Paul, 1979).
7. Ibid.
8. Daniel, *History of Iran*.
9. Ibid.
10. Mary Boyce, trans. *The Letter of Tansar* (Rome: Instituto Italiano per il Medio ed Estremo Oriente, 1968).
11. Richard Frye, *The Heritage of Persia* (Cleveland: World Publishing Co., 1963).
12. Procopius, *History of the Wars*, I, 31, trans. H. B. Dewing (Cambridge, MA: Harvard University Press, 1914).
13. Zachariah of Mytilene, *Syriac Chronicle*, F. J. Hamilton and E. W. Brooks, trans. (London: Methuen, 1899).
14. Frye, *The Heritage of Persia*.
15. Edith Porada, "Iranian Visual Arts: The Art of Sassanians," online article, http://www.iranchamber.com/art/articles/art_of_sassanians.php.
16. Bernard Lewis, *The Middle East* (New York: Scribner, 1995).
17. Victoria Erhart, "The Context and Contents of Priscianus of Lydia's *Solutionum ad Chosroem*," Circle of Ancient Iranian Studies at the School of Oriental and African Studies, http://www.cais-soas.com/CAIS/History/Sasanian/priscianus.htm. Ms. Erhart points out that the assumption that the Academy was formally shut down, while defensible, is built on two somewhat vague comments by John Malalas.
18. Browning, *Justinian and Theodora*.
19. Agathias, *The Histories*, trans. Joseph D. Frendo (New York: De Gruyter, 1975).
20. Massoume Price, "History of Iran: History of Ancient Medicine in Mesopotamia and Iran," online article, http://www.iranchamber.com/history/articles/ancient_medicine_mesopotamia_iran.php.
21. Jamsheed K. Choksky, "Sacral Kingship in Sasanian Iran," Circle of Ancient Iranian Studies at the School of Oriental and African Studies, http://www.caissoas.com/CAIS/History/Sasanian/sacral_kingship.htm.

22. Daniel, *History of Iran.*

23. A. Sh. Shahbazi, "History of Iran: Sassanian Army," online article, http://www.iran chamber.com/history/sassanids/sassanian_army.php.

24. Zeev Rubin, "The Reforms of Khosro Anushirwan," in *The Byzantine and Early Islamic Near East III: States, Resources, and Armies,* Averil Cameron, ed. (Princeton: The Darwin Press, 2004).

25. Richard Frye, *The History of Ancient Iran* (Munich: Beck, 1984).

26. Shahbazi, "Sassanian History."

27. Frye, *The History of Ancient Iran.*

28. Price, "History of Ancient Medicine in Mesopotamia and Iran."

29. Frye, *The History of Ancient Iran.*

30. Porada, "The Art of Sassanians."

31. Frye, *The History of Ancient Iran.*

32. Ibid.

33. Irfan Shahid, "Byzantium and the Arabs in the Fifth Century" (Washington, D.C.: Dumbarton Oaks Research Library, 1989).

34. Barry Hoberman, "The King of Ghassan," *Saudi Aramco World,* vol. 34, no. 2, Spring 1983.

35. Procopius, *Wars,* I, 17.

36. Ibid., VII, 40.

37. Browning, *Justinian and Theodora.*

38. Procopius, *Wars,* II, 10, 4–5.

39. B. H. Liddell Hart, *Strategy* (New York: Plume, 1991).

40. Bury, *History of the Later Roman Empire,* vol. 2.

41. Dionysios Stathakopoulos, *Famine and Pestilence in the Late Roman and Early Byzantine Empire: A Systematic Survey of Subsistence Crises and Epidemics* (London: Ashgate Publishing, 2004).

42. Cyril of Scythiopolis, *Vita* of Kyriakos, quoted in Stathakopoulos, *Famine and Pestilence.*

43. *Vita* of Simeon Stylites the Younger, quoted in Stathakopolus, *Famine and Pestilence.*

44. Graham Twigg, *The Black Death: A Biological Reappraisal* (London: Batsford, 1984).

45. Susan Scott and Christopher Duncan, *Biology of Plagues: Evidence from Historical Populations* (Cambridge: Cambridge University Press, 2001).

46. Michael McCormick, "Towards a Molecular History of the Justinianic Pandemic," *Plague and the End of Antiquity* (Cambridge: Cambridge University Press, forthcoming), cited in Stathakapoulos, *Famine and Pestilence.*

47. Robert Sallares, "Ecology, Evolution, and Early History of Plague," *Plague and the End of Antiquity* (Cambridge: Cambridge University Press, forthcoming), cited in Stathakopoulos, *Famine and Pestilence.*

48. Procopius, *Wars,* II, 24.

49. Procopius, *Wars,* II, 12.

50. Bury, *History of the Later Roman Empire,* vol. 2.

51. Warren Treadgold, *A History of the Byzantine State and Society* (Stanford, CA: Stanford University Press, 1997).

52. Procopius, *Wars,* II, 26.

53. Ibid., II, 27.

54. Ibid., I, 22.
55. Witold Witakowski, *The Chronicle of Pseudo-Dionysius of Tel-Mahre*, Part III (Liverpool: Liverpool University Press, 1997).
56. Ibid.
57. Ibid. The number is, of course, impossible to verify.
58. Michael Dols, *The Black Death in the Middle East* (Princeton: Princeton University Press, 1974), cited in Stathakopoulos, *Famine and Pestilence.*
59. Evagrius Scholasticus, *Ecclesiastical History*, trans. E. Walford (London: H. G. Bohn, 1846).

11. *"No Small Grace"*

1. M. R. Rampino and S. H. Ambrose, "Volcanic Winter in the Garden of Eden: The Toba Super-eruption and the Late Pleistocene Human Population Crash," in Floyd W. McCoy and Grant Heiken, eds., *Volcanic Hazards and Disasters in Human Antiquity* (Boulder, CO: Geological Society of America, 2000).
2. Edward Gibbon, *The Decline and Fall of the Roman Empire*, ch. 10 (New York: E. P. Dutton, 1952).
3. Procopius, *History of the Wars*, V, 12, trans. H. B. Dewing (Cambridge, MA: Harvard University Press, 1914).
4. Agathias, *The Histories*, trans. Joseph D. Frendo (New York: De Gruyter, 1975).
5. Gibbon, *Decline and Fall*, ch. 38.
6. Gregory of Tours: *The History of the Franks*, trans. and introduction by Lewis Thorpe (New York: Penguin Books, 1974).
7. Ibid.
8. Gibbon, *Decline and Fall*, ch. 38.
9. Ibid.
10. Procopius, *Wars*, V, 5.
11. Ibid., V, 5. 8–9.
12. Ibid., VI, 12.
13. Gregory of Tours, *The History of the Franks.*
14. Ibid.
15. Ibid.
16. Josiah C. Russell, "That Earlier Plague," *Demography*, vol. 5, no. 1, 1968.
17. Ibid.
18. Ronald Findlay and Mats Lundahl, "Demographic Shocks and the Factor Proportions Model: From the Plague of Justinian to the Black Death," in Mats Lundahl, Ronald Findlay, Rolf G. H. Henriksson, and Håkan Lindgren, eds., *Eli Heckscher, International Trade and Economic History* (Cambridge, MA: MIT Press, 2006).
19. Stephen Muhlberger, "Overview of Late Antiquity—The Classical Prologue, Section 3: Climate and Economy outside the Mediterranean Basin," ORB Online Encyclopedia, http://www.nipissingu.ca/department/history/MUHLBERGER/ORB/OVC1S3.HTM.
20. Ibid.
21. Peter Brown, *The World in Late Antiquity* (London: Thames & Hudson, 1971).
22. Russell, "That Earlier Plague."
23. William McNeill, *Plagues and Peoples* (New York: Anchor Books, 1998).
24. Findlay and Lundahl, "Demographic Shocks and the Factor Proportions Model."

25. Ibid.

26. Treadgold, *The Byzantine State and Society*.

27. Findlay and Lundahl, "Demographic Shocks and the Factor Proportions Model."

28. Ibid.

29. Ibid.

30. Lynn White, Jr., *Medieval Technology and Social Change* (Oxford: Oxford University Press, 1962); also Marc Bloch, *Feudal Society* (London: Routledge, 1962). Both are quoted in Findlay and Lundahl, "Demographic Shocks and the Factor Proportions Model."

31. Findlay and Lundahl, "Demographic Shocks and the Factor Proportions Model."

32. Ibid.

33. J. R. Maddicott, "Plague in Seventh-Century England," *Past and Present*, no. 156, 1997.

34. Ibid.

35. Ibid.

36. "*Feilire Oengus*," in Reverend P. Power, M.R.I.A., ed., trans., *Life of St. Declan of Ardmore* (London: Irish Texts Society, 1914).

37. Maddicott, "Plague in Seventh-Century England."

38. Ibid.

39. Ibid.

40. Ibid.

41. A. H. M. Jones, *The Later Roman Empire, 284–602* (Baltimore: Johns Hopkins University Press, 1964).

42. McNeill, *Plagues and Peoples*.

43. Maddicott, "Plague in Seventh-Century England."

12. "A Thread You Cannot Unravel"

1. Lawrence Fauber, *Narses: Hammer of the Goths* (New York: St. Martin's, 1990).

2. Procopius, *History of the Wars*, VII, 12, trans. H. B. Dewing (Cambridge, MA: Harvard University Press, 1914).

3. A. H. M. Jones, *The Later Roman Empire, 284–602* (Baltimore: Johns Hopkins University Press, 1964).

4. William Sims Thurman, trans., *Justinian's Edict 6*, "On the Regulation of Skilled Labor," *The Thirteen Edicts of Justinian* (Austin, 1964).

5. Jones, *The Later Roman Empire*.

6. Procopius, *Wars*, VII, 22.

7. C. E. Mallet, "The Empress Theodora," *The English Historical Review*, 1887.

8. J. A. S. Evans, *The Age of Justinian* (New York: Routledge, 2000).

9. Otto G. Von Simson, *Sacred Fortress: Byzantine Art and Statecraft in Ravenna* (Chicago: University of Chicago Press, 1948).

10. Ibid.

11. Antony Bridge, *Theodora: Portrait in a Byzantine Landscape* (London: Cassell, 1978).

12. Ibid.

13. Fauber, *Narses*.

14. John L. Teall, "The Barbarians in Justinian's Armies," *Speculum*, vol. 40, 1965. It should be noted that Teall is unusual in his regard for the Cappadocian, describing

his hand behind a series of reforms intended to provide "honest and efficient provincial administration, thereby enlisting the sympathies of the subject and, hopefully, increasing the tax return ... Certainly John himself enjoyed in popular circles a reputation which his enemies, many among the bureaucracy, were quick to interpret in a pejorative fashion."

15. Victor of Tunnuna, *Chronica*, T. Mommsen, ed., quoted in Dionysios Stathakopoulos, *Famine and Pestilence in the Late Roman and Early Byzantine Empire: A Systematic Survey of Subsistence Crises and Epidemics* (London: Ashgate Publishing, 2004).

16. Witold Witakowski, *The Chronicle of Pseudo-Dionysius of Tel-Mahre*, Part III (Liverpool: Liverpool University Press, 1997).

17. Maurice, *Treatise on Strategy*, trans. George Dennis (Philadelphia: University of Pennsylvania Press, 1984), quoted in Michael Maas, ed., *The Cambridge Companion to the Age of Justinian* (Cambridge: Cambridge University Press, 2005).

18. Fauber, *Narses*.

19. Archibald R. Lewis, *Naval Power and Trade in the Mediterranean*, quoted in Fauber, *Narses*.

20. Procopius, *Wars*, VIII, 23.

21. Ibid.

22. Fauber, *Narses*.

23. Ibid.

24. Bury, *History of the Later Roman Empire*, vol. II.

25. Ibid.

26. H. M. Gwatkin, *Cambridge Medieval History*, vol. 1, quoted in Fauber, *Narses*.

27. M. Kulikowski, "Plague in Spanish Late Antiquity" in *Plague and the End of Antiquity* (Cambridge: Cambridge University Press, forthcoming). Quoted in Stathakopoulos, *Famine and Pestilence*.

28. Sidney Hook, *The Hero in History* (Piscataway, NJ: Transaction Publishers, 1991), quoted in Fauber, *Narses*.

29. Agathias, *The Histories*, trans. Joseph D. Frendo (New York: De Gruyter, 1975).

30. Bury, *History of the Later Roman Empire*, vol. II.

31. Agathias, *Histories*.

32. John of Ephesus, *Evangelical History*, trans. (French) F. Nau, *Revue de l'Orient Chretien* 2, 1897, quoted in Maas, *Age of Justinian*.

33. Evagrius Scholasticus, *Ecclesiastical History*, trans. E. Walford (London: H. G. Bohn, 1846).

34. Ibid.

35. Browning, *Justinian and Theodora* (New York: Praeger, 1971).

36. Procopius, *Wars*, IV, 16.

37. Ibid., IV, 19.4

38. Browning, *Justinian and Theodora*.

39. Procopius, *Wars*, IV, 2.2.

40. Stathakopoulos, *Famine and Pestilence*.

41. Agathias, *Histories*.

42. Clayton O. Jarrett et al., "Flea-borne Transmission Model to Evaluate Vaccine Efficacy against Naturally Acquired Bubonic Plague," *Infection and Immunity*, vol. 72, no. 4, April 2004.

43. Olaf Schneewind et al., "LcV Plague Vaccine with Altered Immunomodulatory Properties," *Infection and Immunity*, vol. 73, no. 8, August 2005. Schneewind is on record as calling *Y. pestis* "the nastiest thing alive."
44. Bury, *History of the Later Roman Empire*, vol. II.
45. Ibid.

13. *"This Country of Silk"*

1. John E. Hill, trans., "The Western Regions according to the *Hou Hanshu*," 2nd edition, online, http://depts.washington.edu/silkroad/texts/hhshu/hou_han_shu.html.
2. William McNeill, *Plagues and Peoples* (New York: Anchor Books, 1976).
3. John Ferguson, "China and Rome," *Aufstieg und Niedergang der Romischen Welt*, Vol., 9.2, 1978.
4. McNeill, *Plagues and Peoples*.
5. Ferguson, "China and Rome."
6. Yong-woo Lee, *Silk Reeling and Testing Manual*, FAO Agricultural Services Bulletin no. 136 (Rome: United Nations, 1999).
7. Manfred Raschke, "New Studies in Roman Commerce with the East," *Aufstieg und Niedergang der Romischen Welt*, vol. 9.2, 1978.
8. Ibid.
9. Ferugson, "China and Rome."
10. Seneca the Younger, *Declamations*, quoted in Ferguson, "China and Rome."
11. Ferguson, "China and Rome."
12. Robert Sabatino Lopez, "Silk Industry in the Byzantine Empire," *Speculum* vol. 20, January 1945.
13. Ibid.
14. Pliny, *Natural History*, quoted in Ferguson, "China and Rome."
15. Ibid.
16. Ferguson, "China and Rome."
17. Ibid.
18. Cosmas Indicopleustes, *Christian Topography*, trans. and ed. with notes and introduction by J. W. McCrindle (Calcultta: The Hakluyt Society, 1897).
19. Procopius, *History of the Wars*, I, 20, trans. H. B. Dewing (Cambridge, MA: Harvard University Press, 1914).
20. A. H. M. Jones, *The Later Roman Empire, 284–602* (Baltimore: Johns Hopkins University Press, 1964).
21. Ibid.
22. Richard Frye, *The History of Ancient Iran* (Munich: Beck, 1984).
23. Procopius, *Wars*, I, 19.
24. Irfan Shahid, "Byzantium and the Arabs in the Fifth Century" (Washington, D.C.: Dumbarton Oaks Research Library, 1989).
25. Frye, *History of Ancient Iran*.
26. Robert Browning, *Justinian and Theodora* (New York: Praeger, 1971).
27. Hill, "The Western Regions according to the *Hou Hanshu*." The date is inferred from a Buddhist history in which the king is identified as Vijaya Dharma whose reign cannot be precisely dated, but certainly covered the year 60 C.E.
28. Ibid.

29. M. A. Stein, "Ancient Khotan," in Manfred Raschke, "New Studies in Roman Commerce with the East," *Aufstieg und Niedergang der Romischen Welt*, vol. 9.2, 1978.
30. Procopius, *Wars*, VIII, 17.
31. R. A. Nicholson, trans., *Ibn Qutaibah*, quoted in Barry Hoberman, "The King of Hassan," *Saudi Aramco World*, March/April 1983.
32. Bernard Lewis, *The Middle East* (New York: Scribner, 1995).
33. L. Conrad, "Epidemic Diseases in Central Syria in the Late Sixth Century: Some New Insights from the Verse of Hassan ibn Thabit," *Byzantine and Modern Greek Studies* vol. 18, 1994.
34. Edward Gibbon, *The Decline and Fall of the Roman Empire*, ch. 50 (New York: E. P. Dutton, 1952).
35. Mark Lewis, Joe Manning, and Walter Scheidel, "The Stanford Ancient Chinese and Mediterranean Empires Comparative History Project" (ACME).
36. Browning, *Justinian and Theodora*.
37. Cyril Mango, "The Triumphal Way of Constantinople and the Golden Gate," *Dumbarton Oaks Papers*, no. 54, 2000.
38. Browning, *Justinian and Theodora*.
39. Paul the Deacon, *Historia Langobardum*, trans. Foulke, quoted in Fauber, *Narses*.
40. Ibid.
41. Browning, *Justinian and Theodora*.
42. Ibid.
43. John Julius Norwich, *A Short History of Byzantium* (New York: Vintage, 1999).
44. Ibid.
45. Dionysios Stathakopoulos, *Famine and Pestilence in the Late Roman and Early Byzantine Empire: A Systematic Survey of Subsistence Crises and Epidemics* (London: Ashgate Publishing, 2004).
46. Ibid.

Epilogue: Yarmuk

1. Chaim Herzog, *The Arab-Israeli Wars* (New York: Random House, 1982).
2. Richard Frye, *The History of Ancient Iran* (Munich: Beck, 1984).
3. P. K. Hitti and F. C. Murgotten, trans., "*The Origins of the Islamic State*, Being a Translation from the Arabic of the *Kitab Futuh al-Buldha* of Ahmad ibn-Jabir al-Baladhuri," *Studies in History, Economics and Public Law*, vol. LXVIII, 1916.
4. David Keys, *Catastrophe* (London: Century, 1999).
5. A. I. Akram, *Sword of Allah: Khalid bin Waleed* (Islamabad, Pakistan: Oxford University Press, 2004).
6. Hitti and Murgotten, "*The Origins of the Islamic State*."
7. K. Gewertz, "Spotlight on the Dark Ages," *Harvard University Gazette*, January 2003.
8. Stathakopoulos, *Famine and Pestilence*.
9. Josiah C. Russell, "That Earlier Plague," *Demography*, vol. 5, no. 1, 1968.

BIBLIOGRAPHICAL NOTE

In the Prologue to *Don Quixote*, Cervantes bemoaned the lack of citations at the end of his book, a deficiency so embarrassing as to stand in the way of the manuscript's completion. And so things remained until a friend advised the author to find another book with the longest bibliography imaginable, and to simply reproduce the same list in the story of the knight of La Mancha, arguing that such an index would give the book a spurious, but surprising, import, and that no one will check anyway.

Readers of *Justinian's Flea* are free to draw their own conclusions, but while hundreds of books and articles have been part of the research for this project (and most of them appear in the book's endnotes) some works, on both the general subject of late antiquity, and on the specific topics that illuminate the period, have been utterly irreplaceable. Anyone seeking to duplicate the excitement that attracted me to the era of *Justinian's Flea* would do well to start at the same place.

If a single predecessor inspired this one, it is Peter Brown's *The World of Late Antiquity*, which may contain the highest wheat-to-chaff ratio of any survey of the period ever attempted . . . though, of course, Professor Brown has the dual advantage of actually knowing everything about his subject and possessing an irritating amount of erudition and style in sharing it. No book by him, on even the narrowest of subjects, can but reward the time spent reading it. Other comprehensive books, by authors no less intimidatingly versed in the period, include Warren Treadgold's *The Byzantine State and Society*; the first volume of Julius Norwich's three-volume history of Byzantium: *The Early Centuries*; Averil Cameron's *The Mediterranean World in Late Antiquity*; and, of course, the relevant chapters of the still immensely readable (and, as readers of *Justinian's Flea* will already have noticed, quotable) *The Decline and Fall of the Roman Empire* by Edward Gibbon and J. B. Bury's

History of the Later Roman Empire. More useful even than Gibbon and Bury were the two volumes of the equally monumental *The Later Roman Empire, 284–602* by Arnold Hugh Martin Jones. The achievement of A. H. M. Jones must, I think, be the only book from which a history of late antiquity could be written without reference to any other.

Two special projects earned a special place in the writing of *Justinian's Flea*: *The Cambridge Companion to the Age of Justinian* and *Readings in Late Antiquity*, both edited by Peter Maas, were illuminating enough on their own, but even more valuable for the light they cast on literally hundreds of other texts, old and new. The same can be said for any of dozens of the monographs produced by the premier Byzantine studies center at the Dumbarton Oaks Research Library and Collection.

On matters of more specific interest, Robert Browning's dual biography *Justinian and Theodora* stands high, as do Herwig Wolfram's *History of the Goths*; Laurence Fauber's biography of Justinian's eunuch-general *Narses: Hammer of the Goths*; *Hagia Sophia* by the architectural historian Rowland J. Mainstone; Tony Honore's definitive biography of the great jurist *Tribonian*; J. A. S. Evans's *Procopius*, and Richard Frye's *History of Ancient Iran*.

Of all the dozens of books written on bubonic plague and corollary subjects, two deserve particular mention; both were extraordinarily helpful: Wendy Orent's *Plague* and Willam McNeill's classic *Plagues and Peoples* . . . almost certainly the best book ever written on the impact of disease upon human history.

This book could, no doubt, have been written with no tools other than those that Gibbon and Bury had to hand, but living in the age of the Internet made it a whole lot easier. Among the literally hundreds of Web sites devoted to the subjects covered herein, I can testify to having spent the most time on these: Paul Halsall's enormous online collections of articles and primary source materials, the Byzantine Studies page (http://www.fordham .edu/halsall/byzantium/index.html) and the Internet Medieval Sourcebook (http://www.fordham.edu/halsall/sbook1b.html); Stephen Muhlberger's ORB Online Encyclopedia, particularly the Overview of Late Antiquity (http:// www.nipissingu.ca/department/history/muhlberger/orb/OVINDEX.htm); and, for an overview of microbiology and epidemiology, Kenneth Todar's Online Encyclopedia of Microbiology (http://www.textbookofbacteriology.net/).

Finally, no bibliographical note could possibly count for much without mentioning, with the sort of gratitude for which words are barely sufficient, the original chroniclers of the plague and its attendant events, including John of Ephesus, Evagrius Scholasticus, and, most important of all, the author of the one absolutely indispensable eyewitness chronicle of Justinian's reign, lawyer and campaigner, amanuensis and scold, the truly indispensable Procopius of Caesarea.

INDEX

Page numbers in *italics* refer to illustrations.